Studies in Computational Intelligence

Volume 995

Series Editor

Janusz Kacprzyk, Polish Academy of Sciences, Warsaw, Poland

The series "Studies in Computational Intelligence" (SCI) publishes new developments and advances in the various areas of computational intelligence—quickly and with a high quality. The intent is to cover the theory, applications, and design methods of computational intelligence, as embedded in the fields of engineering, computer science, physics and life sciences, as well as the methodologies behind them. The series contains monographs, lecture notes and edited volumes in computational intelligence spanning the areas of neural networks, connectionist systems, genetic algorithms, evolutionary computation, artificial intelligence, cellular automata, self-organizing systems, soft computing, fuzzy systems, and hybrid intelligent systems. Of particular value to both the contributors and the readership are the short publication timeframe and the world-wide distribution, which enable both wide and rapid dissemination of research output.

Indexed by SCOPUS, DBLP, WTI Frankfurt eG, zbMATH, SCImago.

All books published in the series are submitted for consideration in Web of Science.

More information about this series at http://www.springer.com/series/7092

Nitul Dutta · Nilesh Jadav · Sudeep Tanwar ·
Hiren Kumar Deva Sarma · Emil Pricop

Cyber Security: Issues and Current Trends

 Springer

Nitul Dutta
Department of Computer Science
and Engineering
SRM University
Mangalagiri, Andhra Pradesh, India

Sudeep Tanwar
Department of Computer Science
and Engineering
Institute of Technology
Nirma University
Ahmedabad, Gujarat, India

Emil Pricop
Automatic Control, Computers
and Electronics Department
Petroleum-Gas University of Ploiesti
Ploieşti, Romania

Nilesh Jadav
Department of Computer Science
and Engineering
Institute of Technology
Nirma University
Ahmedabad, Gujarat, India

Hiren Kumar Deva Sarma
Department of Information Technology
Sikkim Manipal Institute of Technology
Majitar, Sikkim, India

ISSN 1860-949X   ISSN 1860-9503 (electronic)
Studies in Computational Intelligence
ISBN 978-981-16-6596-7   ISBN 978-981-16-6597-4 (eBook)
https://doi.org/10.1007/978-981-16-6597-4

This Springer imprint is published by the registered company Springer Nature Singapore Pte Ltd.
The registered company address is: 152 Beach Road, #21-01/04 Gateway East, Singapore 189721,
Singapore

Preface

In this era of technology, where without the Internet, there is no possibility of doing business, communicating, or assuring e-health and distance learning, it is critical to ensure the security of computer systems and networks. The COVID-19 pandemic showed us that most human activities such as education, work in various fields, shopping, and communication could be done online. The number of attacks trying to compromise the infrastructure has increased proportionally. Obviously, the protection of personal data, privacy, and sometimes anonymity is crucial at that moment. Also, it is a challenging task for the cybersecurity professionals and network administrators to prevent cyberincidents and to investigate if one happened. Moreover, in this rapidly evolving domain, it is essential to stay up to date and to use the latest protection technologies.

The nine chapters of the book comprise many practical examples or short guidelines for using the best tools for anonymity and privacy protection, for analyzing the cybersecurity landscape, increasing security or forensic operations.

Chapter 1 of the book introduces the basic cybersecurity concepts and terminology. Also, it tries to explain the need for cybersecurity in the context of digitalization and increasing Internet usage.

Chapter 2 focuses on the challenging aspects of online anonymity and privacy. The reader will find out the pros and cons of the technologies enabling online anonymity, such as The Onion Router, Invisible Internet Project (I2P), Freenet, and Java Anon Proxy.

The functioning of TOR—The Onion Router—the most well-known privacy browser is presented in Chap. 3 along with the TOR-specific entities. Moreover, in Chap. 4, we discuss the DarkNet and its hidden services. Finally, a simple method for creating a private DarkNet-specific service, a .onion website, is described step by step.

Chapter 5 comprises an extended presentation and various practical examples regarding digital (cyber) forensics used to investigate cyberincidents.

Chapter 6 introduces a powerful cybersecurity protection method, intrusion detection systems, and describes the usage of Snort and OSSEC. Thus, the reader will be able to implement the best solution for their network environment.

Malware analysis, a very challenging and complex task for cybersecurity practitioners and researchers, is presented in Chap. 7.

Chapter 8 has a prominent practical character. We tried to describe creating a cybersecurity laboratory based on virtual machines, honeypots, and other software tools and libraries.

This book concludes with a chapter dedicated to the legal aspects of cybercrimes. We present the legal landscape with respect to cybercrime across various countries in the world and analyze principal cybercrime types and emerging trends.

We believe the contents of the book is a valuable reference regarding the most challenging fields of cybersecurity—preserving the anonymity and privacy of the user, protecting against cyberthreats, and detecting and investigating the attacks or cyber-incidents. These objectives can be reached only by creating awareness on cyberthreats and by educating the home users and the professionals.

Finally, the authors expect this book to be a supportive auxiliary to undergraduate and graduate students, academia, researchers, network administrators, and enthusiasts trying to address cybersecurity issues, cyberforensics, and online problems of privacy and anonymity.

The authors made all efforts to have a good book and hope that interested readers to enjoy reading this book.

Rajkot, India   Nitul Dutta
Rajkot, India   Nilesh Jadav
Ahmedabad, India   Sudeep Tanwar
Majitar, India   Hiren Kumar Deva Sarma
Ploieşti, Romania   Emil Pricop
August 2021

Contents

About the Authors

Nitul Dutta is currently working as an Associate Professor in the Computer Science and Engineering Department at SRM University, Andhra Pradesh. He received his B.E. degree in Computer Science and Engineering from Jorhat Engineering College, Assam and M. Tech. in Information Technology from Tezpur University, Assam. He completed his Ph.D. (Engineering) in the field of Computer Networking from Jadavpur University, West Bengal (2013). He also worked as a Post-Doctoral Fellow (PDF) at Jan Wyzykowski University, Polkowice, Poland. He worked in Tezpur Central University, for nearly nine years, worked at Marwadi University Rajkot, Gujarat and Sikkim Manipal Institute of Technology, Sikkim prior to joining the current assignment. His current research interests are Information Centric Network, mobility management in IPv6 based networks and cognitive radio networks. He is a senior member of IEEE and vice-chair of IEEE Computational Intelligence Society, Gujarat section. He also published more than 30 papers in International journals of repute. He also published two books, Recent Developments on Industrial Control Systems Resilience (Edited), Springer, 2020 and Information Centric Networks (ICN): Architecture & Current Trends, (Reference), Springer, 2021.

Nilesh Jadav is currently pursuing his PhD in Network security from Institute of Technology, Nirma University, Gujarat, India. He completed his Master of Engineering in the field of wireless network from Gujarat Technological University, India. His current research interest includes wireless communication, network traffic analysis and network security.

Sudeep Tanwar (Senior Member, IEEE) is currently working as a Professor with the Computer Science and Engineering Department, Institute of Technology, Nirma University, India. He is also a Visiting Professor with Jan Wyzykowski University, Polkowice, Poland, and the University of Pitesti in Pitesti, Romania. He received B.Tech in 2002 from Kurukshetra University, India, M.Tech (Honor's) in 2009 from Guru Gobind Singh Indraprastha University, Delhi, India and Ph.D. in 2016 with specialization in Wireless Sensor Network from Mewar University, India. He has authored two books and edited 13 books, more than 250 technical articles, including top journals and top conferences, such as IEEE TRANSACTIONS ON NETWORK SCIENCE AND ENGINEERING, IEEE TRANSACTIONS ON VEHICULAR TECHNOLOGY, IEEE TRANSACTIONS ON INDUSTRIAL INFORMATICS, IEEE WIRELESS COMMUNICATIONS, IEEE NETWORKS, ICC, GLOBECOM, and INFOCOM. He initiated the research field of blockchain technology adoption in various verticals, in 2017. His H-index is 43. He actively serves his research communities in various roles. His research interests include blockchain technology, wireless sensor networks, fog computing, smart grid, and the IoT. He is a Final Voting Member of the IEEE ComSoc Tactile Internet Committee, in 2020. He is a Senior Member of IEEE, Member of CSI, IAENG, ISTE, and CSTA, and a member of the Technical Committee on Tactile Internet of IEEE Communication Society. He has been awarded the Best Research Paper Awards from IEEE IWCMC-2021, IEEE GLOBECOM 2018, IEEE ICC 2019, and Springer ICRIC-2019. He has served many international conferences as a member of the Organizing Committee, such as the Publication Chair for FTNCT-2020, ICCIC 2020, and WiMob2019, a member of the Advisory Board for ICACCT-2021 and ICACI 2020, a Workshop Co-Chair for CIS 2021, and a General Chair for IC4S 2019, 2020, and ICCSDF 2020. He is also serving the editorial boards of *Physical Communication, Computer Communications, International Journal of Communication System*, and *Security and Privacy*. He is also leading the ST Research Laboratory, where group members are working on the latest cutting-edge technologies.

Hiren Kumar Deva Sarma is currently a professor in the Department of Information Technology, Sikkim Manipal Institute of Technology, Sikkim. He received a B.E. degree in Mechanical Engineering from Assam Engineering College in 1998. He completed Master of Technology in Information Technology from Tezpur University in 2000. He received his Ph.D. degree from Jadavpur University (Department of Computer Science and Engineering) in 2013. He is a recipient of Young Scientist Award from International Union of Radio Science (URSI) awarded in the XVIII General Assembly 2005, held at New Delhi, India. Dr. Sarma received IEEE Early Adopter Award in the year 2014. He has published more than seventy research papers in different International Journals, referred International and National Conferences of repute. His current research interests are wireless sensor networks, mobility management in IPv6-based network, cognitive radio networks, ICN, network security, robotics, distributed computing and big data analytics.

Emil Pricop is currently an Associate Professor and the Head of the Automatic Control, Computers and Electronics Department of the Petroleum-Gas University of Ploiesti, Romania. Also, he is an invited professor at the Computer Engineering Department of Faculty of Engineering (FoE), Marwadi University, Rajkot, Gujarat, India. He has held the position of Senior Lecturer since 2018. Dr. Pricop is teaching computer networking, software engineering, human-computer interaction, and critical infrastructure protection courses. He received his Ph.D. in Systems Engineering from Petroleum-Gas University of Ploiesti by defending in May 2017 the thesis

"Research regarding the security of control systems." His research interest is cyber-security, focusing primarily on industrial control systems security. Dr. Emil Pricop is co-editor of two books published by Springer, namely Recent Advances in Systems Safety & Security (Springer, 2016) and Recent Developments on Industrial Control Systems Resilience (Springer, 2020). Also, Dr. Pricop is the author or co-author of 2 national (Romanian) patents, six (6) book chapters published in books edited by Springer and over 30 papers in journals or international conferences. Since 2013, Dr. Pricop is the initiator and chairman of the International Workshop on Systems Safety and Security – IWSSS, a prestigious scientific event organized annually. Dr. Pricop participated in more than 100 technical program committees of prestigious international conferences organized under the auspices of IEEE. He has held the vice-chair position of the IEEE Young Professionals Affinity Group - Romania Section from 2017 to 2019.

Chapter 1
Introduction to Cybersecurity

1 Introduction to Cybersecurity

1.1 Introduction

Cybersecurity primarily deals with a clear interpretation of various issues related to attacks that occur on digital resources and subsequent remedies therein. While discussing the safety against attacks or crimes against digital resources, the confidentiality, integrity, and availability of such information must be preserved. In the last few decades, society has become incredibly dependent on communication networks and digital information. As a consequence, cyberattacks are gaining considerable attention among intruders. Moreover, many researchers in cyberspace used available tools and tactics for unlawful access and manipulation of digital content. Hence, such digital crimes are nowadays a nightmare for individuals as well as the society at large. At the same time, the rate of cyberattacks and crimes is increasing as it is easier to perform than physical attacks. The only things required by a criminal are a computer and an Internet connection. The free availability of tools and software for the purpose makes the hackers more convenient to exploit.

Moreover, unconstrained geography and distance in accessing data make it more convenient for attackers because, in such situations, it is difficult to reveal the criminal's identity and prosecute due to anonymity. It is expected that the rate of cybercrime and sophistication in attacks will grow in the coming time. Therefore, it is essential to understand the cybersecurity concerns of every individual.

1.2 The Necessity of Cybersecurity

In the recent past, society has witnessed a tremendous technology-reliant service than ever before and it is expected to grow more in the coming time. Nowadays, people prefer to store their valuable files and data on open platforms such as Google

N. Dutta et al., *Cyber Security: Issues and Current Trends*, Studies in Computational Intelligence 995, https://doi.org/10.1007/978-981-16-6597-4_1

Drive. They share their documents through Dropbox, etc. One way these facilities make our life digitally comfortable but impress a higher risk in data leaks that could result in theft of identity as well as sensitive information. In fact, cloud storage is one of the prime targets of a security breach that may lead to revealing information like social security numbers, credit card information, or bank account details.

Various governments agencies over the globe are trying hard to bring awareness about cybercrimes as well as to protect against data breaches. The general data protection regulation (GDPR), which is applicable in Europe, is one of the prominent examples. Governments in Europe have made it mandatory for all organizations that operate in the EU to:

- Communicate any occurrences of data breaches;
- Have a data protection officer (DPO) in every organization;
- Prior consent before for personal information processing;
- Preserve anonymity in the data processing.

This shift of public disclosure is not limited to Europe but has been formulated in various countries. For example, in 2003, California was among the first state in the USA to oversee data breach disclosure which has the following mandate:

- Requirement to notify about the data breach as soon as possible;
- Government should know about the breaches;
- Fines should be imposed.

This brings to an establishment of standard boards such as the National Institute of Standards and Technology [1] (NIST). They create standards and framework that helps companies audit their security infrastructure and prevent any cyberattack. They update their frameworks every three years, where it is necessary for an organization to update them from their end. Multiple threat actors play a significant role in a successful cyberattack. Data theft [2] is an overpriced threat and hard to stop from exponential growth. Contrary, cybercriminals have simple attacks which are in the most accessible form to enter into the system resources. From there, they can spread their affect to perform more sophisticated hybrid attacks. Therefore, there is a need for better cybersecurity to keep our data and resources safe by implementing and applying security protocols and standards into the organization. Most importantly, there should be an awareness among the employees with respect to security.

1.3 Cybersecurity and Ethics

The broad meaning of "ethics" in the context of life is *"how to live a good life."* Humans always try to live good, for that they take better opportunities. The ethics [3, 4] measure whether the opportunities are better or not. With technological advancement, human beings have several options to seek for a good life. The speed and scale of the technical advances bring humans to either take a good way or the wrong way to get a better life. Numerous examples can be seen as a "lack of ethics." For example—companies are collecting data from biometrics, face-recognition to track users. There is no consistent rule or framework for collecting such data, and, perhaps, there is a risk of leaking data into public. The data can have user medical history, credit card information, or other sensitive information that intruders can use to pursue nationwide massive attacks. Another such incident happens with companies such as Facebook, Google, Amazon, and Microsoft, where they collect huge information from our daily searches. Such data is often leaked into the public, a relevant example being Facebook Cambridge Analytics (2017). They are plenty such incident happened in the past, still, the basic rule of ethics is not applied in technology, and hence, it is important to understand and aware people about it. The simple rule of ethics with respect to security—"Do not do something wrong in a cyberworld where others have to pay in everyday life". There are no general rules specifying ethics; however, it is a moral responsibility of a user not to do:

- Use of any malicious software;
- Use of cyberbully;
- Out hears the communication line (passive eavesdropper);
- Using someone else's password for your benefit;
- Follow the copyright restriction while downloading movies, games, and software (adhere to license).

There are ethical issues in cybersecurity which in turn leads to harm for some users and benefit for some users. The person has a choice—"ethical" or "unethical" to achieve a good life. It depends on what he chose, if he chooses ethics—he is using good ways to get a benefit, or else harming others to get benefitted. Table 1—shows the ethical issues in cybersecurity.

These issues can be rectified by awareness programs, cultivating ethical teachings among employees and students. In addition, the institutions that provide cybersecurity as a professional course should mandatorily teach cybersecurity ethics. Hence, we can ensure that future security professionals stay on the right side.

2 Domains of Cybersecurity

In earlier days, if we wanted to defend our nation, we had to apply strict defense line across the country. An attacker can attack from any place by crossing the defense line. Hence, the guardians need to place a stronger borderline to prevent this breaching. Modern-day attacking strategies [5] revolve around the same concept, where there are several defenses such as firewalls, detection systems, antivirus, and access control. An attacker has to surpass all these obstacles to breach into the system. Therefore, a security professional has to work in several domains to gather knowledge about it and prevent it. However, it will be challenging to accumulate all knowledge in one place; hence, we keep all this knowledge in separate cybersecurity domains. Moreover, we expect each domain to have a professional who can manage it when

Table 1 Cybersecurity ethical issues in terms of privacy, property, resource allocation, and data breach disclosure with its effect on the organization

Ethical issues	Effect on the organization
Ethical issue in privacy	• Identity theft • Gather sensitive information for blackmailing, extortion • Big companies such as Facebook and Google have control over our personal information
Ethical issue in property	• With cyberintrusion, one can check electronic funds, steal intellectual property • Not only an individual gets harmed, but sometimes the nation has to pay—attack such as *Stuxnet* on control system is entirely unethical, which can lead to loss of lives
Ethical issue in resource allocation	Happens due to a lack of decision. For example—Your company provides low-level security solutions. However, you risk giving an efficient detection system or a better Incident response team; however, both are lacking in the company
Ethical issue in disclosure	Lack of transparency when disclosing a bug or vulnerability. It is essential to have a timely notification about the bug disclosed to the customer to patch it before something terrible happens

an attack happens. To discuss about all fields of cybersecurity is out of the scope of this book; hence, we are only considering the essential domains such as,

(a)   *Computer Security*—is the protection of the computer systems from unauthorized access. The primary function is to update and patch the standalone machine. Further, it has hardware, firmware, and software components to protect from vulnerabilities—backdoors, malware, and privilege escalation [6]. Computers are the most attacked entity due to their openness to network, installation of any software, and use of any peripherals. The attacker has a broad scope to attack such a device which has these many open doors. Computer security professional [7] has below responsibilities:

- Integrate IT specifications to audit risks;
- Design efficient security measures and data recovery plans;
- Configure and update security software;
- Monitor and analyze network activity to identify an intrusion;
- Action plan on privacy breaches;
- Should conduct expert training for awareness program;
- Inspect any hardware or firmware-related vulnerability.

A vulnerability is a weakness or loophole in a computer or connected network which can be exploited by the attacker to damage or to manipulate the system. *Network vulnerabilities*—weakness in the hardware or software network which can leverage a possible intrusion. *Operating system vulnerabilities*—vulnerabilities related to the operating system, such as poorly configured access control or account management that can cause damage to the system. Perhaps, an attacker can install a hidden backdoor or perform privilege escalation to access the system or its files. *Human vulnerabilities*—where the human can make mistakes such as exposed login credentials of superuser may install malware unknowingly or creating a weak password.

(b) *Data Security*—The twenty-first century is the century of data, where data is an essential commodity for economists, societies, and democracies. With big data coming into the picture, the importance of data gets twofold; later by applying artificial intelligence with big data, the organization can make intelligent decisions and transform their business. However, with such a craze of data, there are limitations, such as massive data thefts and breaches that have occurred recently. The USA is the most affected country by data breaches, damage cost up to $8 million. Several data security issues might be faced by the organization [8], for instance, accidentally share sensitive data. It is a common practice in an organization where an employee accidentally shares, grants access, or mishandles critical data. On the other hand, with the help of social engineering, phishing is the foremost lethal attack to collect data from users. It will manipulate the human mind in accordance with *greed, curiosity, empathy,* and *fear.* This will trick the user's mind into clicking on malicious links, which leads to a successful phishing attack. There is awareness about phishing among users, but still, this attack has its own strategy to become fortunate. Insider threat is another category of people, where an employee from the organization itself becomes an attacker and threatens data security. Cloud storage is not much secured due to the lack of stable configuration, insecure interfaces, and DoS attacks. Hence, there is a need for data security experts who can take care of such challenges. Data security is the process of protecting critical data from unauthorized access. Data security professionals have below job responsibilities:

- Should know the data (type of data, where it is stored, and use of data);
- Should create data masking—no real data, type of data remains same, but the values get changed;
- Should be able to install data protection software (DPS);
- Report security breaches;
- Should be able to perform rigorous testing—penetration testing, to gather flaws of the system;
- Should be applying governance, risk, and compliance (GRC) to secure the data;
- The organization should conduct data security audits every few months;
- Should be able to provide strong authentication and authorization methods.

(c) *Network Security*—A network consists of several devices interconnected with each other using wired cable or wirelessly with some addressing mechanism (IP and MAC). In a broader term, the network can be made up of several technologies such as wired (devices are connected with cables), wireless (devices are connected with access points), or cellular connection (devices are connected wirelessly with base stations). Open System Interconnection (OSI) consists of seven layers (physical, data link, network, transport, session, presentation, and application layer) that form standards for feasible communication between networks. Regarding OSI, each layer has its own security challenges, which need to be taken care of [9] as presented in Table 2. For example, cryptography can help upper layers secure confidentiality and integrity, but researchers have not explored lower layers too much until now (Table 2).

Today's Internet and its under relying technologies keep upgrading, bringing new frameworks and platforms to develop more modern technologies such as artificial intelligence, 5G data networks, blockchain, and quantum computers. But, perhaps, these advancements are duplex. Attackers are also sitting on the same Internet, analyzing and monitoring the recent trends. In this way, they are also advancing their attacking strategies to attack more severely. Attacks, for instance—distributed denial of service (DDoS), man in the middle (MiTM), unauthorized access, injection attacks, eavesdropping, jamming, and malware attacks are possible on the network.

On top of that, anonymous Internet is currently in recent trends, so the attackers can use such private Internet to proliferate their attacks. When private connections are used, it is a challenging task for a network administrator to track the attacker. Moreover, some connections use layered encryption and relays, which are completely

Table 2 OSI layer (top to bottom approach) with their relevant protocols and possible attacks on each layer

OSI layers (top to bottom)	Basic protocols used	Attacks
Application layer	DNS, SMTP, HTTPS, Telnet	Malware, SMTP attack, DNS tunneling, FTP bounce
Presentation layer/presentation layer	SSL, HTTP/HTML, SSH, TLS, Apple filling (AFP)	Information retrieval, remote procedure call attacks, NetBIOS attacks
Transport layer	TCP, UDP, stream control transmission (SCTP), fiber channel (FCP), encapsulating security (ESP), authentication over IP	TCP/UDP attacks, SYN flood, port attacks, TCP sequence prediction attacks
Network layer	NAT, IP, ICMP, RIP, ARP, OSPF	Identity theft, IP spoofing, sybil attacks, selective forwarding hijacking attack, unauthorized access
Data link layer	LLDP, ATM, Cisco discovery (CDP), IEE 802.xx, HDLC, NDP	ARP spoof, MAC flood, spanning tree attack
Physical layer	USB, ISDN, Etherloop, DSL, ARINC, Frame relay	Eavesdropper, interference attack, jamming, impersonate attack. floods, open wall ports

decentralized and anonymous, therefore, hard to track. For security solutions, as said, encryption can help a lot in the upper layer; however, with modern digital processors and quantum computers, it is a simple task to decode the encrypted message. Therefore, a solution such as embedded hardware authentication, blockchain cryptography, physical layer security, intelligent firewalls, and detection mechanism is needed to implement.

The network security engineers have below mentioned job responsibilities:

- Identify, monitor, and define the requirement of the overall security of the system;
- Installation and configuration of detection and prevention system;
- Should be able to create efficient rules for firewalls;
- Should be able to create and configure access control list for secure authorization of resources;
- Inspect the system for vulnerabilities in software and hardware;
- Should have a comprehensive knowledge of network protocols and their configuration;
- Should be able to install security infrastructure devices;
- Should be able to investigate threats from the network traffic;
- Maintenance of switches and servers.

(d) *Digital Forensics*—is a field used to investigate digital evidence (data) collected or retrieved from cybercrime. The digital evidence can be in any form—hard drives, disk, USBs, or any other storage media. The objective of the forensic investigator is to perform rigorous research on the evidence to find the real culprit. This branch is sometimes also incorporated in the incident response team, where the ultimate goal is to secure the organization's assets from the attacker. Forensics is a vast subject to explore, and due to this fact, it calls a large number of diverse people to work with each other. For instance, a network engineer can perform log analysis and yield information about how the attacker accessed the network and what resources he had examined or modified. Evidence is vital in CyberForensics. Due to the volatility of the storage media, preserving the evidence is the first challenge. Secondly, gathering and collecting information from the evidence is an arduous task as evidence may be secured with encryption, and decrypting it without the key is tough. However, there are high chances of data loss. In order to prove the crime in a court of law, an investigator tries to find a hidden folder, unallocated disk space for deleted or damaged files from the evidence.

Forensics investigator has following responsibilities:

- Should be able to run any forensic tool to extract and analyze data (FTK, Encase, etc.);
- Should be able to recover damaged, deleted, protected, or encrypted files;
- Collection of volatile data in an acceptable way;
- Should be able to perform a vulnerability assessment;
- Collaborate with reverse engineers and incident response team for speedy investigation;

Table 3 Domains of cybersecurity 2021, from the well-known certification—CISSP by (ISC)[2]

Domains of cybersecurity	It covers
Identity and access management	Secure access control, authorization, identity management, authentication, physical and logical access to assets
Security and risk management	CIA, compliance requirements, IT policies, risk management, security governance requirements
Asset security	Privacy, data security, retention periods, inspect ownership of information
Security architecture and engineering	Design physical security, cryptosystem, audit and mitigating system vulnerabilities, concept of security models, securing information system
Communication and network security	Design secure network architectures, secure network devices and assets, secure communication channels, monitoring network traffic
Security assessment and testing	Designing test strategies, validating assessment, securing control testing, inspect output, internal and external audits
Security operations	Log analysis, incident management, recovery plans, managing physical security, securing resources, efficient investigations
Software development security	Software development lifecycle, software security, firm guidelines, and standards for coding

- Should have analytical, critical, and problem-solving skills.

Many non-profit organizations, such as International Information System Security Certification Consortium (ISC)[2], provide cybersecurity training and industry certification. One of the renowned certificates [10]—*Certified Information Systems Security Professional (CISSP),* has been taken as a benchmark to define cybersecurity domains. These are the widely accepted domain of cybersecurity in 2021 (Table 3).

Though we have secure solutions available in every domain, attackers are still finding minuscule vulnerabilities from the system to damage the organization's reputation or for their own benefit. By keeping up-to-date with the latest security risk, the company can implement an effective cybersecurity strategy to defend ourselves from the harmful breach.

3   Threats and Actors

The scope of this book is to make you understand the fundamentals of cybersecurity, where there are many terms that, at first, are not easy to understand. This section covers the essential threats and threats actors that reside in the cybersecurity domain and play both negative and positive roles.

3.1 Threats in Cyberspace

A threat is a negative scenario that leads to an inevitable outcome like damage, loss, or harm, not only to business owners but also to home users. It is an activity performed by the malicious technocrat intended to comprise confidentiality, integrity, and availability of an information system [11]. This is a concerning topic by many security professionals, as every day, the threat landscape is growing, getting competent, and more challenging. For example, Cloudflare—a free and open-source content delivery network had recently fixed a critical bug (path traversal) which affects 12.7% of the website on the Internet [12]. Microsoft found another vulnerability of local privilege escalation (unpatched), affecting the Windows Print Spooler. When Windows Print Spooler service is not performing well, an attacker can use this opportunity to run an arbitrary code with SYSTEM privileges, where he can install programs and perform superuser activity [13]. On the other hand, ransomware is the biggest challenge and threat which directly affects system resources. Without any ransom paid, decrypting the file is not possible, and one can be in locked condition either he can wait to get the key from the attacker once the ransom is paid (not reliable) or simply format the system drives (data loss).

3.2 Types of Threats

Each day is possible to encounter a different threat in any part of the world, and the severity of each threat is increasing day by day. The known threats, such as phishing, DoS, MiTM, malware, and SQL injection, can be encountered using upper layer security solutions. However, the biggest challenge to tackle is zero-day vulnerability and hybrid attacks since we don't have a possible security solution for them. This subsection will analyze the context of the 2020 year and the new cybersecurity challenges to individuals and enterprises.

(a) *Cloud Vulnerability*—Due to COVID-19, firms went remotely; they operate their file sharing and storage on the cloud. In the early stage of COVID-19, there was a hustle of lockdowns; hence, many firms have not correctly configured their servers, network, and system resources. Due to that, the attackers had this great opportunity to lurk and attack. Adapting cloud is a possible solution for remote firms, but not applying security policies or misconfiguration in the cloud may lead to a potential risk (Fig. 1).

(b) *AI and Machine Learning*—AI and its domains have been largely used to automate different tasks. In cybersecurity, it has been used to create intelligent solutions—intelligent firewalls, antivirus, or detection systems. Mainly, it is used for automated threat intelligence, to analyze massive incoming attacks, and to secure data. Perhaps, AI and machine learning go both ways, with attackers using the same domain to automate their attacks. For instance, creating data

Which of the following represents the biggest cloud security challenges for your organization?

(Percent of respondents, N=456, five responses accepted, seven most frequently reported challenges shown)

Maintaining secure configurations for our cloud-resident workloads	39%
Satisfying our security team that our public cloud infrastructure is secure	38%
Maintaining strong and consistent security across our own data center and public cloud environments in use	38%
Cloud-related security event management challenges	37%
Aligning regulatory compliance requirements with my organization's cloud strategy	30%
Inability for existing network-security controls to provide visibility into public cloud-resident workloads	30%
Inability to automate the application of security controls due to the lack of integration with DevOps tools	30%

Fig. 1 Cloud security: One of the security challenges of 2021, among others. Misconfiguration and unauthorized access of the cloud are the highest percentage (38%) in cloud security [14]

poisoning, model stealing, and AI fuzzing methods to automate attacks can optimize the attack methodology.

Data privacy in machine learning is an important issue, as machine learning approaches are dependent on the dataset quality. Columns in the dataset represent features (name, age, gender, etc.), and rows represent features value. Attackers are interested in rows where there is an actual value. Dataset has sensitive values such as—medical history, power plants threshold values, and network information, which an attacker can use to damage the infrastructure or harm the user.

(c) *Social Engineering Attacks* are lethal but straightforward attacks where users can be tricked to provide sensitive information, resulting in massive data breaches. Mostly, it is used in scamming businesses. However, most companies are applying secure spam models and block such phishing attacks. However, due to emotions (greedy, curiosity, empathy, and fear), users get easily trapped in the attack and release crucial information.

Deepfake is a recently developed mechanism to create fake videos or audios, especially to create riots or to spread wrong information. Hackers use AI-based technology to create counterfeit videos by swapping people's faces and modifying their speech (Fig. 2).

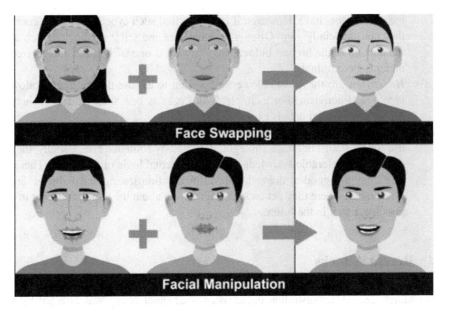

Fig. 2 Deepfake: upper image shows us face swapping, where attackers have merged a boy and a girl image to create an individual. The bottom image shows facial manipulation, where two characters are exchanging their physique to create a completely different person [15]

3.3 Threat Actors and Types of Threat Actors

An important component in the threat landscape is the "threat actor." It can be anyone who has the capability to influence IT security. Precisely, a threat actor is an individual or a group of people intended to carry out cybercrimes or attacks but not limited to. It can be an organization or nation-state involved in the process. It is a crucial step to know these actors who are playing a role in attacking your infrastructure. Knowing them allows the cybersecurity personnel to create better security policies to protect the systems.

(a) *Cyberterrorist* is a widespread global threat actor plagued in most countries. They are focused on damaging critical services and causing harm to humans. Their main targets are business, control systems, and IT infrastructure.

(b) *Government/state-sponsored actors* are often funded or sponsored by nation-states to collect sensitive information and exfiltrate secure property. The motivation behind such an attack is either the political, military or commercial interest of their county. Their strategies are not much active. Therefore, they work passively, like an organization is a target, they get into the organization (by being an employee), work for the long term, and silently pass the sensitive information to their real personnel.

(c) *Cybercriminals* are the general criminals, not very much targeted and operate on a broad mass of victims. They do not have any defined strategies to attack as the other actors have. However, it is hard to find such cybercriminals because they are financially gain. Often, they steal data and sell it on the dark web or via auction to the highest bidder. Ransomware is one such attack categorized under cybercriminals.

(d) *Hacktivists* are the whistleblowers who want to expose the truth and follow ideological activism. For example, WikiLeaks is one such platform where highly confidential data is exposed to the public.

(e) *Script Kiddies* call them hackers. However, they lack skill and knowledge. They use scripts, tools, or attack strategies designed by someone else. Usually, they don't have preferable knowledge behind the actual logic of the attack. Therefore, they are good at doing low-level penetration testing and vulnerability assessment, where they get enough information about the target, resulting in a possible attack in the future.

4 Recent Attacks

On March 28, 2021, Australian broadcaster—"Channel Nine" was affected by a cyberattack. Hackers rendered the channel into GIFs, unable to air that Sunday news and other events. Officials declared this attack as cyberespionage, but they didn't take any name of the country or state but believe countries such as Russia, China, and Iran may be behind this attack [16].

Harris Federation, a London-based non-profit organization in the education sector, was attacked by ransomware. Due to this incident, the institution has to temporarily disable its e-mail system along with other devices. This results in, unable to access coursework and other correspondence by the 37,000 students [17]. In response to the attack, they declared the following statement "we are committed to ensuring the UK education sector is resilient against cyberthreats and have published practical resources to help establishments improve their cybersecurity and response to cyberincidents."

Another ransomware attack occurred on March 21, 2021. This time, the victim was one of the biggest cyberinsurance companies, namely CNA financial [18]. The attack was so intense that the company had to shut down service for three consecutive days to prevent further compromise. They immediately called forensic experts and law enforcement to investigate and minimize the data breach. A new version of Phoenix malware (cryptolocker) has been used in ransomware.

Sometimes, attacks on the control systems lead to considerable damage in the country. A relevant example is a ransomware attack on Florida-based water system that shocked the world. Cybercriminals have attempted to poison the water supply by increasing the amount of sodium hydroxide. They remotely accessed the water supply and managed to increase the level of sodium hydroxide to a dangerous level. Fortunately, this comes to the notice by the supervisor on his computer screen, and appropriate actions have been taken immediately before the water leaves to the city's water supply.

In February 2021, Accellion, a secure file sharing company, was affected by a cyberattack. Attackers exploited a zero-day vulnerability in its file transfer appliance (FTA), which the company has been using for the last 20 years. Unfortunately, hackers have learned the technology and exploited not one but three vulnerabilities in FTA. All vulnerabilities are patched now and listed as CVE-2021–27,101, CVE-2021–27,102, CVE-2021–27,103, and CVE-2021–27,104.

There are tons of other cyberattacks and, more severe, possibly happening today while writing this book, or in the future when you will read this book. The important thing is to get aware and updated on these attacks. The biggest weakness of any organization is their unaware employees. Therefore, there should be some mandatory cybersecurity awareness programs to understand and get knowledge about recent cyberthreats and their preventive measures. It has been observed that social engineering attack is the most common and significant attack for data breaches. In order to deal it with such attacks, the companies should implement phishing incident response tool such as Threat Alert Button (TAB). In addition to that, we should periodically conduct vulnerability assessment and penetration testing. This is important as it will reveal any unknown or known vulnerability which the attacker could exploit. On the last note, always keep all the systems up-to-date and patched to protect from any cyberattack.

5 Awareness of Cybersecurity in Educational System

In this era of technology, where without the Internet, there is no possibility of doing business, communicating, or assuring e-health and distance learning it is critical to ensure the security of computer systems and networks. It is evident that the Internet is not only used by adults, and due to the COVID pandemic, the education sector has used remote learning as a medium to continue children's education. Though it has vast potential and benefits, excessive use can lead to cyberaddiction [19, 20]. Not only that, cybercriminals can use children and youth to encourage in doing cyberattacks. Also, children can become the victims of cybercrime. Many guardians are unaware of their children's activity in cyberspace. As a result, the children may get harassed, molested, or sexually assaulted on social media or any other similar platforms. There were days when these children were concerned about their homework or outdoor sports competition. Unfortunately, due to the COVID pandemic, they are now sitting in one corner of the room with their gadgets, connected to the Internet and various social networking platforms. The parents are also responsible for the same since they give these devices in the form of rewards such as good grades in an exam or birthday present. This way, an adolescent gets addicted to the Internet, which has a diversity of information. Obviously, it depends on the child what he is searching and using on the Internet. Moreover, there is the risk that the children live in two lives—real life and virtual life, which is nowadays dominating. Perhaps, it is the responsibility of the parent to look after the child's Internet activity, thus taking good care of the child's real life.

Chatroulette is an online random chatroom to meet girls, guys, comedians, musicians, etc. Analysis from RJmetric [21] shows us that there are 70% of the user are having age between 20 and 30, 23% are teens and youngers aged between 14 and 18. This data is of one such site, but there are plenty of websites available where children and youngsters spend their days and nights, for example, Omegle, another website for random chat and videos. Due to the pandemic, the traffic increased from 35 million visitors to 65 million visitors a month. The traffic is mainly from the USA, Mexico, and India. The site claims to be moderated and age verified, but in actuality , there are children aged 13 or 14 years doing an illicit activity. An investigation by BBC got dozens of men aged under 18, some are below 10, doing explicit things, and this raised a severe concern about children's future.

Therefore, a cybersecurity education program is needed in schools to educate on both importance and negative impact of the Internet. There are many benefits if a school has a curriculum for cybersecurity education. Otherwise, a student has to pay money or time on several seminars and webinars for a short duration. Though seminars are valuable, where you can interact with other individuals like you, such programs lack activity and assessment [22]. A school can become a knowledge center where teachers can discuss with students and organize several activities related to cybersecurity awareness. To change the mindset of an individual, it is vital that the program must be adequately assessed; hence, a fruitful outcome can come. Primary schools have challenges as they have students from a very early age, who knows the Internet but not its underlying principles. They know how to search and find the results from search engines but do not know the difference between safe web links and phishing web links. Their teachers have to work on; they can use videos or cartoons to make them aware and understand cybersecurity topics. In addition, the school should repeatedly organize speech competitions, essay writing, debates, and discussion in the classroom to enhance their knowledge regarding the subject [23].

6 The Outline of the Book

The second chapter of the book "Being Hidden and Anonymous" is focused on challenging aspects regarding online privacy. Nowadays, the need for anonymity is obvious when almost all online activities are recorded and analyzed by various companies and governmental institutions for different purposes. Therefore, the authors examine the fundamental aspects of online security concepts: confidentiality, integrity, and availability. Also, the technologies enabling online anonymity and privacy, such as The Onion Router, Invisible Internet Project (I2P), Freenet, and Java Anon Proxy, are presented exhaustively, and their pros and cons are compared.

The Onion Router (TOR) is widely used for ensuring online privacy and for limiting government surveillance. However, it is sometimes used as a gateway for accessing the dark Internet, where significant illegal activities occur. In addition, TOR can be a valuable resource for communicating sensitive information.

The third chapter presents the functioning of TOR and its routing protocols. The authors describe the TOR-specific entities extensively and show some methods for analyzing the current TOR network status. The mobile client for TOR—Orbot—is introduced along with TOR network usage guidelines.

The fourth chapter of the book discusses exciting aspects regarding DarkNet and its hidden services. The Internet is currently providing more than the surface web, the well-known websites accessible through search engines such as Google, Bing, Yahoo, or Yandex. The Deep Web, which accounts for about 90% of the websites, comprises financial records, academic databases, scientific reports, which requires advanced search skills to be accessed. The DarkNet, estimated to account for about 6% percent of the available resources on the Internet, can be only accessed using special tools such as TOR, I2P, or Freenet. In this chapter, the authors focus on the DarNet access methodology and on creating your own private DarkNet-specific resources.

Cybersecurity is a complex task. We should protect from various attacks and threats, but we have to analyze and investigate the intrusions, attack tracks, and even the events on DarkNet. Digital forensics is the method used to investigate cybercrimes, and the fifth chapter focuses on an extended presentation of this complex process. The authors present a variety of practical examples on computer and network forensics.

The sixth chapter introduces a powerful cybersecurity protection method, intrusion detection systems. A short technology review is presented, and the characteristics of other security tools such as firewalls are emphasized. The core of that chapter is the characterization of IDS and a thorough presentation of Snort usage. An alternative solution—open-source host-based intrusion detection system (OSSEC) is also presented. Thus, the reader will be able to implement the best solution for their network environment.

Cybersecurity cannot be assured without knowing very well the behavior of the threats. Therefore, the seventh chapter of this book focuses on malware analysis, a very challenging and complex task for cybersecurity practitioners and researchers. First, the authors present the main malware categories and their symptoms. Then, the focus of the chapter is on a detailed description of the malware detection systems and analysis methods.

The cybersecurity fundamental and expert notions introduced in the previous chapters can be studied in a well-configured virtual laboratory environment. Therefore, the eighth chapter of the book has a particular practical focus on the creation and configuration of a virtual machine-based laboratory for security study. In addition, a large number of software tools and libraries are discussed. Also, the authors present a case study for intrusion detection using honeypot tools.

The book concludes with a chapter dedicated to the legal aspects of cybercrimes. The legal landscape regarding cybercrime across various countries in the world is analyzed in detail.

References

1. "National institute of standards and technology|NIST", NIST (2021). [Online]. Available: https://www.nist.gov/
2. L.E. Conde-Zhingre, B.D. Piedra-Cevallos, G.I. Cueva-Alvarado, R.F. Espinosa-Espinosa, Cybersecurity as a protection factor in the development of Smart Cities. 2020 15th Iberian conference on information systems and technologies (CISTI) (2020), pp. 1–5
3. P. Formosa, M. Wilson, D. Richards, A principlist framework for cybersecurity ethics. Comput. Secur. **109**, 102382 (2021)
4. K. Macnish, J. van der Ham, Ethics in cybersecurity research and practice. Technol. Soc. **63**, 101382 (2020)
5. K. Fowler, Cybersecurity. *Enterprise Risk Management* (2016), pp. 91–108
6. S. Van Till, All security is now cybersecurity. *The Five Technological Forces Disrupting Security* (2018), pp. 97–106
7. C. Sennewald, C. Baillie, Computers and effective security management, *Effective Security Management* (2021), pp. 245–260
8. L. Sun, H. Zhang, C. Fang, Data security governance in the era of big data: status, challenges, and prospects. Data Science and Management **2**, 41–44 (2021)
9. P. Pandya, Local area network security *Network and System Security* (2014), pp. 259–290
10. 2021. [Online]. Available: https://www.isc2.org/Certifications/CISSP
11. M. Guitton, Facing cyberthreats: answering the new security challenges of the digital age. Comput. Hum. Behav. **95**, 175–176 (2019)
12. "RCE vulnerability in cloudflare CDN could have allowed complete compromise of websites", The daily swig|cybersecurity news and views (2021). [Online]. Available: https://portswigger.net/daily-swig/rce-vulnerability-in-cloudflare-cdn-could-have-allowed-complete-compromise-of-websites
13. "Security update guide—microsoft security response center", msrc.microsoft.com (2021). [Online]. Available: https://msrc.microsoft.com/updateguide/vulnerability/CVE-2021-34527.
14. M. Davidson, *oracle.com* (2021). [Online]. Available: https://www.oracle.com/a/ocom/docs/dc/final-oracle-and-kpmg-cloud-threat-report2019.pdf?elqTrackId=063c9f4a2a5b465ab55b734007a900f0&elqaid=79797&elqat=2.
15. K. Howard, Deconstructing deepfakes—how do they work and what are the risks? *WatchBlog: official blog of the U.S. government accountability office* (2021). [Online]. Available: https://blog.gao.gov/2020/10/20/deconstructing-deepfakes-how-do-they-work-and-what-are-the-risks/
16. "Cyber-attack disrupts live broadcasts by channel nine", Abc.net.au (2021). [Online]. Available: https://www.abc.net.au/news/2021-03-28/channel-9-off-air-due-to-technical-issues/100034364
17. Itpro.co.uk (2021). [Online]. Available: https://www.itpro.co.uk/security/ransomware/359064/harris-federation-ransomware-attack
18. K. Mehrotra, W. Turton, "Bloomberg—are you a robot?", Bloomberg.com (2021). [Online]. Available: https://www.bloomberg.com/news/articles/2021-05-20/cna-financial-paid-40-million-in-ransom-after-march-cyberattack.
19. Y. Hong, S. Furnell, Understanding cybersecurity behavioral habits: Insights from situational support. J. Inf. Secur. Appl. **57**, 102710 (2021)
20. K. Cabaj, D. Domingos, Z. Kotulski, A. Respício, Cybersecurity education: evolution of the discipline and analysis of master programs. Comput. Secur. **75**, 24–35 (2018)
21. "Chatroulette is 89 percent Male, 47 percent American, and 13 percent perverts", *Tinyurl.com* (2021). [Online]. Available: https://tinyurl.com/56k64bjd
22. F. Quayyum, D. Cruzes, L. Jaccheri, Cybersecurity awareness for children: a systematic literature review. Int. J. Child-Comput. Interact. **30**, 100343 (2021)
23. V. Švábenský, P. Čeleda, J. Vykopal, S. Brišáková, Cybersecurity knowledge and skills taught in capture the flag challenges. Comput. Secur. **102**, 102154 (2021)

Chapter 2
Being Hidden and Anonymous

1 Introduction

The idea of anonymity and privacy of data over the Internet has been changed to a greater extent, and the reason behind this is a revolution of technology. It not only upgrades the deprecated computers but comes up with a series of advancements in automation and telecommunication. This advancement in technology brings a lot of freedom to the users to save their personal information. However, those users may not be aware that their data may be used for illicit acts. A survey on the type, volume, and depth of personal information saved on the Internet is carried out by Pew Research, as presented in reference [1]. The report states that 66% of users upload their photos online, 50% save their date of birth on social media platforms, and 46% save their e-mail addresses. In addition to that, they also save their company profile, home addresses, cell number, and sometimes video of their own. The survey also reveals that most users are afraid of hackers and advertisers. Most of them are unaware of accessing data by any third-party services, government, and law enforcement companies. Needless to say, this information is enough for any intruder to enter into someone's system and get access to everything.

Two diagrams from reference [1] are presented in Figs. 1 and 2 in order to give readers an overview of the private information stored on the Internet. Figure 1 shows the private data that is uploaded and stored online by the users. Results from Fig. 1 show us the percentage amount of data stored during their web surfing sessions. Hackers may use this data to gain access to users' systems using different techniques available for the purpose. Figure 2 shows who are the potential users of the private data.

N. Dutta et al., *Cyber Security: Issues and Current Trends*, Studies in Computational Intelligence 995, https://doi.org/10.1007/978-981-16-6597-4_2

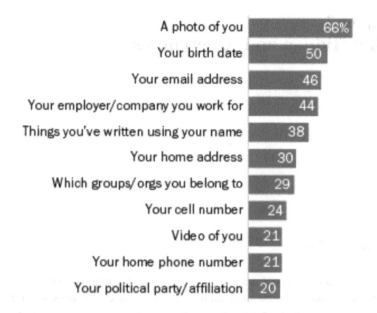

Fig. 1 Types of data stored online by users. Results show in percent the amount of data stored during their web surfing sessions. Hackers may use these data to gain access to their system using different techniques available for the purpose. *Source* Survey Conducted by Pew Research 2014—pewresearch.org [2]

1.1   The Need for Anonymity

Being alert and avoiding exposure of personal as well as other sensitive information to anyone is a prime necessity while storing data or working online. Every day we store our small portion of personal information on the cloud, which has a constant risk of data disclosure, and hence there is a privacy concern and needs to be taken care of. In addition, people are changing their perspective from normal Internet to anonymous Internet in order to hide their identity. From Pew Research [4], 86% of the Internet users are removing or masking their digital footprints, started using VPNs, or are masking their IP address.

In order to be able to discuss anonymity, we should discuss the first three fundamental concepts regarding cybersecurity, namely the *Confidentiality, Integrity, and Availability (CIA)* triad. It is an important pillar of online security and privacy, giving users some methods that can protect their data and identity from data breaches and identity theft [5].

Confidentiality: It is important to secure private and sensitive information from any outliers in the digital world. Confidentiality restricts the access to information for any unauthorized person. It ensures the data that is passed from sender to receiver cannot be read and understood by any third party (unauthorized person). For example, if you sign up to the Facebook page, then the request–response between you and the server

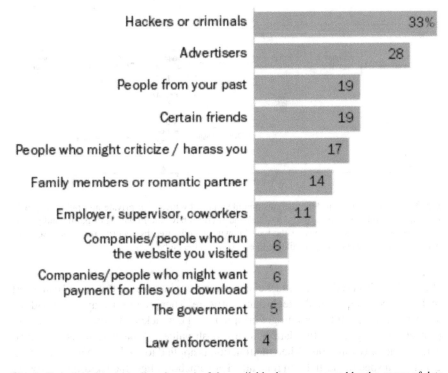

Fig. 2 Potential users (actually misusers) of the available data as supposed by the owner of data who stored them online. People mostly assume that advertisers and hackers use such data available online. However, there is a possibility that such data is used by the government and law enforcement departments as well. *Source* Survey Conducted by Pew Research 2014—pewresearch.org [3]

should be confidential, and no other third party can come in between to access that information. Information such as health records, financial accounts, criminal records, source code, trade secrets, and military tactical plans can be considered confidential information. Any disclosure of such information can result in heavy loss for an individual, a corporation, or even for a nation-state. To ensure data confidentiality, powerful encryption algorithms were developed. Encryption involves coding a message using a specially created encoding algorithm and an encryption key (being shared or asymmetric—public and private key pair). The message is encoded using the key and can be deciphered only using the same key, in the case of symmetric encryption, or using the corresponding public key, in the case of asymmetric encryption. A general flow of asymmetric encryption for protecting message confidentiality is shown in Fig. 3.

Integrity: It is a vital component of the triad, where it deals with the protection against the alteration/modification of data when it passes from a user to another user. It ensures the data that the sender sends is exactly the same received by the destination. It measures the accuracy, correctness, and completeness of the data. This could be very helpful in restricting active attacks such as man-in-the-middle

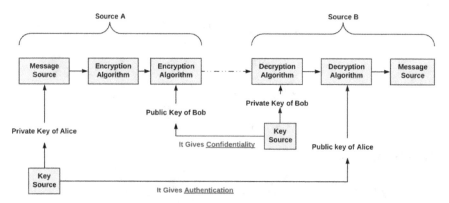

Fig. 3 Flow diagram that demonstrates confidentiality by means of private and public key pair. Descriptive caption required. Part A (Alice) and B (Bob) have their own set of public and private key pairs derived by using any of the well-known methods used by security protocols. Both the parties make their public keys available to everyone on the globe

attacks. It is observed that an attacker can reverse engineer the software and can insert malicious content inside the legitimate software. Then, the user may download the software from the Internet and get crippled by the attacker. A possible solution to protect the integrity of the data is to use hash values. A hash value is generated using a one-way function. When the function is applied to the legitimate software, it generates a specific hash. Even if an attacker somehow alters the legitimate software by reverse engineering, the resulting hash value will not be the same. We can compare the hash value on both sides and observe that the software is legitimate or not. For example, we have a file named "test.txt" containing a simple text string "This is a test document" we calculate the hash value and save it, as shown in Fig. 4. Furthermore,

Data Format:	Data:			Data Format:	Data:		
File ▼		test.txt	...	File ▼		b:\test.txt	...
	Key Format:	Key:			Key Format:	Key:	
☐ HMAC	Text string ▼			☐ HMAC	Text string ▼		
☑ MD5	0f6181daa8ff29c67a755b2a6b0d8b8e			☑ MD5	9e6d9644f36fc8aef1e96da9435ca260		
☐ MD4				☐ MD4			
☑ SHA1	fb522d544850baa90aa8a61b69e830f4ae19b69c			☑ SHA1	228a63dc93795c876f14f461714a936f7721cd27		
☐ SHA256				☐ SHA256			
☐ SHA384				☐ SHA384			
☐ SHA512				☐ SHA512			
☑ RIPEMD160	d82412cdaaa37dd1230cbf6b3cd9c1d84b6e0263			☑ RIPEMD160	5e5ea2962557513aa2b45fd2b868a0b0016ffa80		
☐ PANAMA				☐ PANAMA			
☐ TIGER				☐ TIGER			

Fig. 4 A scenario to demonstrate assurance of integrity by generating a *hash* value of file contents. Even if the content of any document is altered by a single bit, the hash value of both the files (original and the altered one) is changed significantly. Thus, it enables the user to check if the received file is tampered with or not

if we add some text string to the same document "This is another test string", because of the alteration of some bytes, we can see there is a change in the hash value, and we can easily know that some had changed the content.

Availability: Information security also takes care of the availability of the data to the authorized user. It ensures that the software or service a user intends to use is available whenever the user requests it. It should perform hardware and software repair immediately and maintains the functioning of the operating system. Also, it should upgrade the system whenever it is required. To make availability possible, it is necessary to have a strong recovery plan and an effective fault tolerance mechanism. There are high chances of non-availability of services in the communication networks because of not having adequate bandwidth, data losses, and even caused by a distributed denial-of-service (DDoS) attack. We can store a backup copy of that data to some other geographically isolated location to resolve data loss. To defend from DDoS attacks, we can use a potent firewall and effective firewall rules and proxy servers to block unauthorized requests. The CIA triad plays a crucial role because all defensive measures, security controls and mechanisms, that we implement to achieve one or more of these protection types.

2 The Onion Router

Many anonymous networks that proliferated over the Internet claim to preserve our identity and not disclose our private information. The Onion Router (TOR) [6] is one of the anonymous and encrypted networks. For its functioning, it uses the Onion Routing (OR) protocol. The motivation behind Onion Routing, which has a "low-latency Internet-based connection" is to protect the user from traffic analysis, eavesdropping, and other attacks by outsiders and insiders. The OR uses hidden routers which are hosted inside the TOR network. These routers are completely anonymous, and no one knows where they are physically placed. TOR network operates with almost 7000 dedicated Onion Routers separated worldwide, generating almost 400Gbps for advertised bandwidth, as shown in Fig. 5. To connect to any particular server, OR uses a combination of three or six routers called "relay circuits" from a pool of routers. Each router knows only two things: the next router's IP address and a decryption key, which greatly helps make the TOR network a complete anonymous network. In reference [7], the authors show a comprehensive overview of the TOR network and its technical methods: proxy, VPN, and Onion Routing. It also specifies the actual configuration of the TOR network, varieties of user mistakes while using the service, and several specific technical design issues.

TOR can save your real identity from third-party adversaries; however, it is very susceptible to traffic analysis. An experienced attacker or networking expert can analyze the traffic and can reveal what content is going inside the network.

The TOR network comprises a wide range of legal and illegal hidden services deployed by hosts whose identity is anonymous too. Reference [8] gives significant information about the TOR network as it surveyed TOR data communication,

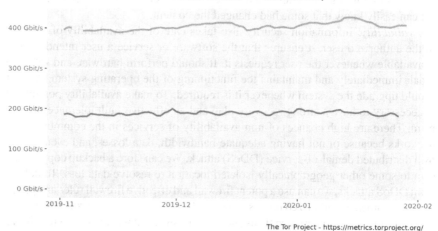

Fig. 5 Bandwidth of the TOR network as advertised by the TOR project. The project owner claims that it operates with almost 7000 dedicated Onion Routers separated worldwide, generating almost 400 Gbps

providing common and existing techniques in data communication to understand and avoid traffic analysis as well as data integrity of the TOR network. The authors of Refs. [9, 10] have specified a browser-based attack where the user browser can be tricked to send a distinctive signal over the TOR network. The signal can be further analyzed using traffic analysis to break the anonymity. It also shows how exit nodes and their policies can be broken to simplify traffic analysis (Fig. 6).

The author of Ref. [11] created a very inspiring work of several attacks on the TOR network. This paper provides an interesting attacks taxonomy that will help in understanding their functioning. Despite all these loopholes, TOR is among the top prior network in the category of anonymous networks. Due to the increase in relays, bandwidth, and access to any service, people continuously add to the network and hide their identities.

3 Invisible Internet Project (IIP or I2P)

Invisible Internet Project is a beta project, started in 2003, implemented as a mixed network that allows peer–peer anonymous communication [12]. It builds with the same motivation as the TOR network. However, it is far better than the TOR network, as it provides more anonymity than TOR. The authors of Ref. [13] present an empirical study of the I2P network, where they measure network properties such as population, churn rate, router type, and I2P peers. These anonymous systems are based on directory systems, either centralized or decentralized. In Ref. [14], the authors

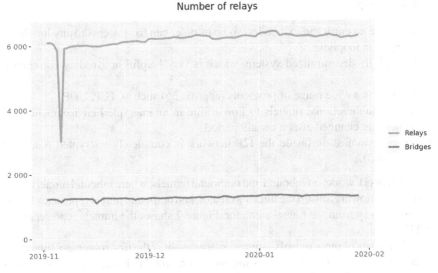

Fig. 6 Count of relays and bridges used in the TOR network over time. The diagram shows the data from 2019 to 2020. There are approx. 6000 relays and 1400 bridges in the TOR network are reported in various surveys

evaluate the performance of the I2P network and its decentralized directory "netDB", comparing it with Kademlia (KAD) design, it turns out that the former version has KAD distributed hash table vulnerabilities, from which authors learned and improved the netDB design.

Pros of Tor:

- It is the first network categorized as an anonymous network worldwide.
- The services are completely hidden and hosted by an anonymous host.
- It is a large-scale network and still continuously improvising.

Cons of Tor:

- There are entry and exit nodes that can be exploited using various mechanisms as these nodes are at a low level of encryption.
- It is a highly centralized system. Ten directory servers keep looking for performance, reliability of nodes, and other statistics within the network. Any attackers aim to gain access to these directory servers to get the plethora of information about TOR users.
- Even though there here are larger numbers of users, we have only 8000 nodes. Therefore, TOR is asymmetric in nature.

Pros of I2P:

- There are no entry and exit nodes, so no intruder can have access to any low-level encryption loophole.
- It is a fully decentralized system, which is very helpful in avoiding correlation attacks.
- It supports a wide range of protocols (approx. 56) such as TCP, UDP.
- It uses unidirectional tunnels to flow traffic in an encrypted environment. The tunnels get changed after a certain period.
- All communication inside the I2P network is completely encrypted (end–end encrypted).

I2P network works on inbound and outbound tunnels, where inbound tunnel represents traffic flowing toward the tunnel's creator, and outbound tunnel represents traffic that goes away from the tunnel's creator. Figure 7 shows the tunnels representation in I2P.

I2P consists of three tunnels—tunnel gateway—it is the first router in a tunnel for User 1 and User 2. The gateway is the originating router (Routers 1 and 4), tunnel endpoint is the last router in a tunnel (Routers 3 and 6). Thus, tunnel participants are all routers in the I2P network, except for gateway and endpoint (Routers 2 and 5), as presented in Fig. 8.

When User 1 wants to send a message to User 2, it sends it to the outbound tunnel, and the reply comes back from User 2 to User 1 in the inbound tunnel. It definitely hides the content, but anyone can know that you are using I2P. The tunnels that I2P uses are completely unidirectional and not bidirectional. It maintains a network database (NetDB) that contains *routerinfo* and *leasesets*, where *routerinfo* provides transport addresses, public–private keys, relevant information about the network, and *leasesets* give information about any particular destination the user wants to contact. NetDB is a custom structured distributed hash table (DHT), created by modifying the Kademlia algorithm, in order to find the inbound tunnel efficiently.

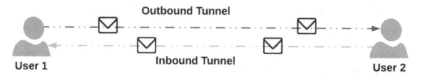

Fig. 7 An example of tunnel representation in I2P. Here User 1 is the tunnel's creator, and User 2 is the intended target of User 1, also called tunnel endpoint

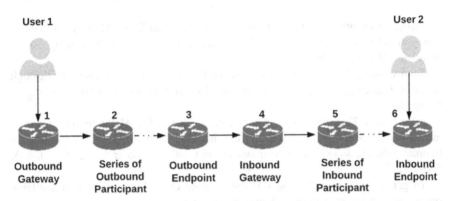

Fig. 8 A sequence of routers used in I2P. There are six different routers used in the tunnel formation between the tunnel owner and the endpoint. Such a series of routers in the tunnel adequately ensures anonymity

3.1 Working of I2P

I2P network comprises several sets of routers (nodes) connected with a number of unidirectional inbound and outbound tunnels. Each router can be identified using a cryptographic router identity and communicate with each other using protocols such as TCP and UDP. I2P clients can send and receive messages using their own cryptographic identifiers. The client is authorized to connect to any router and lease a temporary allocation of these inbound and outbound tunnels (virtual paths), used for sending and receiving the messages. The message routed between client and server is garlic wrapped, using three layers of encryption, as described below and presented in Fig. 9.

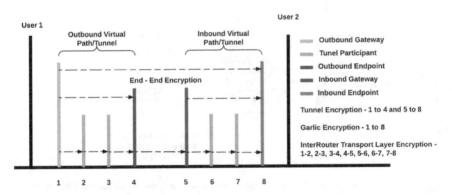

Fig. 9 Working of Invisible Internet Project (I2P). There are eight different entities by which the packet moves from sender to receiver using three types of encryption. These encryption layers make anonymity stronger that preserves the user's privacy

Garlic encryption: This encryption uses ElGamal/SessionTag + Advanced Encryption Standard (AES), to verify the successful delivery of the message to the recipient. (Router 1 to Router 7).

Tunnel Encryption: It uses private key AES to encrypt messages flowing through the tunnel gateway to the tunnel endpoint. (Router 1 to Router 4 and Router 5 to Router 8).

Inter Router Transport Layer Encryption: It uses Diffie–Hellman key exchange and Station-to-Station protocol (to provide encryption between each router) and AES to protect the message flowing from one router to other. (Router 1–2, 2–3, 3–4, 4–5, 5–6, 6–7, 7–8).

To get more information on I2P, the detailed official documentation available online on https://geti2p.net/en/ should be consulted.

We have downloaded and installed I2P on a host running Microsoft Windows 8 operating system. For further usage, I2P should be started (Start I2P restartable), and it will configure certain settings for the I2P router, such as bandwidth test and trackers. After the successful configuration, the user is ready to surf I2P services with an URL (127.0.0.1:7657/home). To use I2P "hidden services", you need to find specially designed websites, called "eepsites". They have an extension (domain). i2p, which is similar to the "TOR Hidden Services" .onion extension.

Mostly, all services are listed on the configuration page itself, and the detailed services are listed below.

- **Hidden Services and Web Browsing**—You can search any eepsite having an extension. i2p, default i2p pages can be found on 127.0.0.1/console as "I2P hidden service of interest". *Leywork.i2p* is an effective search engine for only the I2P network. It never indexes any. onion or surface Internet website. Another service *Ransack.i2p* also gives results for both surface Internet and eepsites. You can find more services from—*seeker.i2p, i2pwiki* and *identiguy.i2p*. The i2p URL structure is shown in Fig. 10.

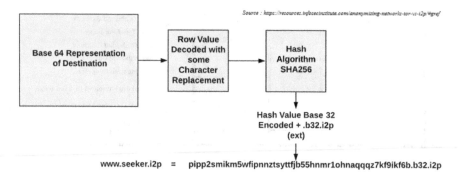

Source : https://resources.infosecinstitute.com/anonymizing-networks-tor-vs-i2p/#gref

| Base 64 Representation of Destination | Row Value Decoded with some Character Replacement | Hash Algorithm SHA256 |

Hash Value Base 32 Encoded + .b32.i2p (ext)

www.seeker.i2p = pipp2smikm5wfipnnztsyttfjb55hnmr1ohnaqqqz7kf9ikf6b.b32.i2p

Fig. 10 Representation of an eepsite link having an *i2p* extension. These are not regular domains (.com, .in, .net), and they will not reveal the actual domain, as it is encoded and hashed with different algorithms, to protect the real owner of the website

- **Anonymous mail**—They have an anonymous mail server known as *i2p-bote*—an internal, serverless, and end–end encrypted mail server. Moreover, it also has inbuilt mail—*susimail*, to not disguise your mail identity. An effort has been made to protect your IP address in the TOR network as it is very susceptible to such protocols where e-mail can be taken as an attack vector to reveal your identity. However, I2P mail is completely anonymous and encrypted.
- **Anonymous file transfer**—*i2psnark* is an integrated I2P service for torrent file transfer, and it provides anonymous encrypted bit torrent transfer.
- **Anonymous chat**—You can use Internet Relay Chat (IRC) by setting up an IRC client application as "*Hexcat*" and configure it to user 127.0.0.1:6668 as their server. I2P comes with an IRC server hosted by the postman, echelon, and zI0, so your message gets directed to any of these servers and received by the intended receiver. All communication is encrypted to prevent any intruder interception.
- **Forums and blogging**—Blogging is one of the features that you really want to do in a hidden environment, as you are writing against the adversaries. To become anonymous, you need to keep your software and application up-to-date (Apache, MySQL, Python, PHP).

The I2P runs in any browser configured with the correct settings. However, it is not feasible, as a normal browser can fingerprint too much information and share it with intermediary and third-party services. In order to defend against this threat, we can put our I2P router in the TOR browser and make it hardcore to restrict any data leakage.

I2P installation on Microsoft Windows operating system is an effortless task. However, Linux and Debian systems are more protective in nature that is the reason they are more secure and hard to attack.

In the following paragraphs, we will describe the installation process of I2P on a Kali Linux host along with the TOR browser and the configuration of the I2P router to work with the TOR browser. It is recommended that you add a new user to Kali System, as I2P never starts with a root user. Therefore, we need to add one non-root user, and you can do that by using the command *adduser test*, setting the password, adding the newly created test user to the sudoers file, and switching to *test* user, execute this command one by one to install I2P router.

```
$ adduser test
$ usermod –a –G sudo test
```

We have to add repositories to /etc./apt/sources.list.d/i2p.list according to your Debian distribution, you can check our Debian version using—/etc./debian_version. Execute these commands for Kali Rolling,

```
$ sudo mousepad /etc/apt/sources.list.d/i2p.list
```

```
deb https://deb.i2p2.de/ stretch main
deb-src https://deb.i2p2.de/ stretch main
```

```
$ sudo apt-get install i2p
```

Once you imported the repository, you have to sign these repositories and you have to check the fingerprint and owner of the key.

```
$ curl -o i2p-debian-repo.key.asc https://geti2p.net/_static/i2p-debian-repo.key.asc
$ gpg -n --import --import-options import-show i2p-debian-repo.key.asc
```

Add the key to APT keyring, and update it for fetching the latest list of packages from the repository, then install i2p-keyring.

```
$ sudo apt-key add i2p-debian-repo.key.asc
$ sudo apt-get update
$ sudo apt-get install i2p i2p-keyring
```

Start your i2prouter (it will not run if you start I2P as the root user, make sure you create a new non-root user).

```
$ i2prouter start
```

It will start in your default browser with an URL 127.0.0.1:7657. The initial steps of the I2P router are for configuring and testing the network. You can set your bandwidth precisely to your network connection. In addition to that, update and refresh your peers, it should be greater than 10, otherwise, your connection to open any I2P service is going to slow. The running phase of the I2P service is presented in Fig. 11. Moreover, it also shows exploratory tunnels created by your router to communicate eepsites and other services. These exploratory tunnels consist of gateway, participant, and endpoint, as shown in Fig. 9.

I2P will start with your default browser, but we need to open it with the TOR browser. Therefore, we need to configure this default setting by going to—http://127.0.0.1:7657/configservice and find "*Launch Browser on Router Startup*"—change it to—"*Do not view console on startup*".

To set up the I2P network in the TOR browser, you need to install the TOR browser according to your compatible Linux distribution. In our case, it is Kali Linux and we can install it by using the below command (remember, you are with test user).

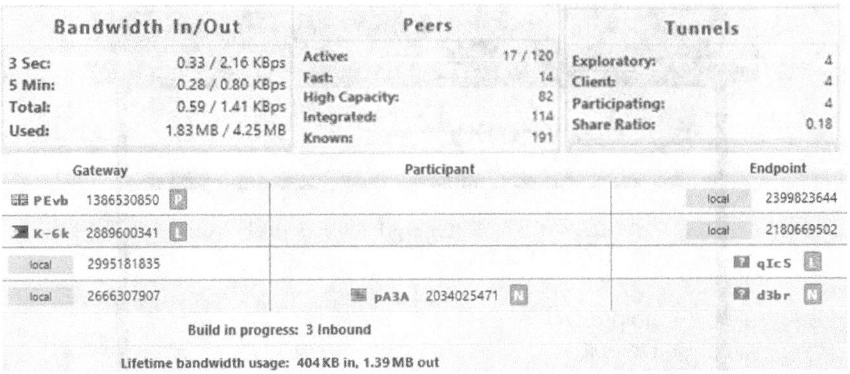

Bandwidth In/Out		Peers		Tunnels	
3 Sec:	0.33 / 2.16 KBps	Active:	17 / 120	Exploratory:	4
5 Min:	0.28 / 0.80 KBps	Fast:	14	Client:	4
Total:	0.59 / 1.41 KBps	High Capacity:	82	Participating:	4
Used:	1.83 MB / 4.25 MB	Integrated:	114	Share Ratio:	0.18
		Known:	191		

Gateway			Participant			Endpoint	
PEvb	1386530850					local	2399823644
K-6k	2889600341					local	2180669502
local	2995181835					qIcS	
local	2666307907		pA3A	2034025471		d3br	

Build in progress: 3 Inbound

Lifetime bandwidth usage: 404 KB in, 1.39 MB out

Fig. 11 Running phase of the I2P network. We can test and configure the network according to the network status, such as bandwidth, tunnels, and peers. Active peers should be more than 10, otherwise, you will get a slow connection to open any I2P website

$ wget https://www.torproject.org/dist/torbrowser/9.0.4/tor-browser-linux64-9.0.4_en-US.tar.xz

$ tar –xvJf tor-browser-linux64-9.0.4_en-US.tar.xz

The following step is to set up our I2P proxy so that the I2P network opens up in the TOR browser and not with the default browser. The reason we use TOR is to become completely anonymous and not to leave any backtrack path that can reveal our identity. However, the method is very tidy as we need to change proxy settings again and again in our browser. To solve this issue, we need an add-on named "FoxyProxy" that makes our work easier. FoxyProxy is a proxy setter that provides a platform to change our proxy with a single click. Download it from—https://addons.mozilla.org/en-US/firefox/addon/foxyproxy-standard/, and once installed, it shows up in the menu bar, as shown in Fig. 12.

FoxyProxy gives efficient management of your proxies. Specifically for the I2P router, TinHat had written down a configuration file for FoxyProxy, consisting of a rule set to handle different traffic connecting to different services such as TOR, I2P, Freenet, and so on. You can get the foxyproxy.xml file from https://thetinhat.com/tutorials/darknets/foxyproxy.xml. On Linux systems, you can use the following command:

$ wget https://thetinhat.com/tutorials/darknets/foxyproxy.xml

Fig. 12 FoxyProxy add-on for browser—there is a need for a proxy to use the I2P network. 127.0.0.1:4444 is the I2P proxy from which we can connect to the I2P network

The configuration file consists of different proxies, such as 127.0.0.1:7657 (for I2P router). 127.0.0.1:4444 (connect to I2P websites—eepsites), 127.0.0.1:9150 (connect to TOR network), 127.0.0.1:8888 (connect to Freenet). With a single click, you can flip your proxy. To start the I2P router, connect to 127.0.0.1:7657 and open it in the TOR browser, and you will get the I2P Router Console page, as shown in Fig. 13.

You also have to delete the proxy of 127.0.0.1:9150 (for TOR network) as now we don't want to direct our traffic from the TOR network, but instead, we want to use only the I2P network. Therefore, either delete the proxy or chose the option "Direct (No Proxy)". To become safer on the TOR browser, we have the option to select the "security level" as required. The recommended level is "Safest" from Standard, Safer, and Safest, which will disable JavaScript, Flash and some icons, images, and math symbols. There will surely be a degradation in performance, but you can get a proper anonymous channel to communicate with different Internet services.

4   Freenet

Freenet is another anonymous service that will protect your identity from mass surveillance, trackers, and law enforcement bodies [15]. It is a peer–peer service for censorship-resistant communication that uses a decentralized system to store and retrieve data from the users. Freenet can be downloaded from (https://freenetproject.org/pages/download.html for any compatible operating system.

Install and start Freenet, and it will fire up the default browser pointing to http://www.localhost:8888. You can then select the "security level"—low security, high

I2P ROUTER CONSOLE

Version: 0.9.44-0-3
Uptime: 68 min

Bandwidth In/out

3 Sec: 0.07 / 0.15 KBps
5 Min: 0.31 / 0.63 KBps
Total: 0.41 / 0.95 KBps
Used: 1.57 MB / 3.75 MB

Network: Firewalled

Local Tunnels

shared clients
shared clients (DSA)

Dec 1, 2019 0.9.44 RELEASED

0.9.44 contains an important fix for a denial of service issue in hidden services handling possible.

The release includes initial support for new end-to-end encryption (proposal 144). Work are changes to the console home page, and new embedded HTML5 media players included. Tunnel build fixes should result in faster startup for some users.

As usual, we recommend that you update to this release. The best way to maintain secu

News last updated 69 days ago. News last checked 75 min ago.

Applications

Addressbook   Email   Hidden Services Manager

Fig. 13 I2P Router Console Page. It will show various configuration settings for Address book, Hidden Service Manager, Anonymous Git Hosting to achieve proper anonymity. The page will also show the status of bandwidth, security level, and tunnels

security, and custom security, followed by asking the datastore size and bandwidth configuration. Freenet provides an HTTP interface for browsing websites and it only accesses the content which has previously been inserted/hosted in the Freenet network. The connection works as peer–peer, as your node is connected to another node and the node connected to another and so on, your data request will reach the data holder by directing from node to node, as presented in Fig. 14. All such nodes are also using Freenet and therefore, it is slightly not secure compared to TOR and I2P. Even though the traffic is encrypted, there is still a risk in revealing the identity, so the recommendation is to connect to only the people whom you know.

There are directories of websites available on Freenet, as shown in Fig. 15.

One thing that hinders all these anonymous services is their dependence on platforms or operating systems. For example, Freenet is built with Java, and an attacker can try to find vulnerabilities and bugs of java and then use these bugs or vulnerabilities to exploit Freenet and such services.

Freenet is vulnerable to denial-of-service (DoS) and spam attacks. Still, no possible vulnerabilities are found over the Internet, apart from one— bypass something—CVE-2019–9673 https://www.cvedetails.com/vulnerability-list/vendor_id-19946/product_id-55582/year-2019/Freenetproject-Freenet.html, as shown in Fig. 16.

Due to its design strengths: decentralize network, dynamic routing, and high resilience to attacks, Freenet is a very adaptable network to use.

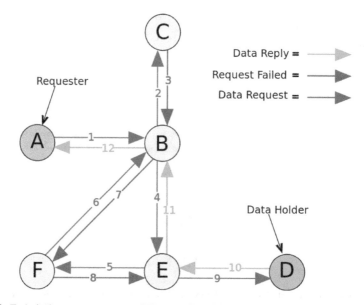

Fig. 14 Technical message request of Freenet. The message request moves from one node to another and follows heuristic routing, where routing is based on the node which serves the key faster. The request continuously moves and fails at 3 (dead end), 7 (a loop), and eventually reaches the data holder. *Source* Wikiwand.com [16]

Fig. 15 Directories of Websites on Freenet. It shows various indexes—Enzo's—most Freenet websites, Nerdageddon—Annoying content removed, and Filtered Index

Year	# of Vulnerabilities	DoS	Code Execution	Overflow	Memory Corruption	Sql Injection	XSS
2019	1						
Total	1						
% Of All		0.0	0.0	0.0	0.0	0.0	0.0

Directory Traversal	Http Response Splitting	Bypass something	Gain Information	Gain Privileges	CSRF	File Inclusion	# of exploits
		1					
		1					
0.0	0.0	100.0	0.0	0.0	0.0	0.0	

Fig. 16 Common Vulnerabilities and Exposure (CVE)-2019–9673. This will exploit Freenet 1483, which has a MIME type bypass, that will allow JavaScript execution

5 Java Anon Proxy (JAP)

JonDonymn or Java Anon Proxy (JAP) is a similar proxy service to Freenet. It is an open-source service based on the Java platform. There is a premium account available for JAP that gives more flexibility in order to configure ports and other settings.

To use JAP, it is highly recommended to use JonDoBrowser, an anonymous Web browser similar to the TOR browser. You can get it from https://anonymous-proxy-ser vers.net/en/jondofox.html, which allows you to surf the Internet with pseudonymity. After installing, it will give you an assistant to configure initial settings for anonymity test, bandwidth, and proxy. The main advantage of JAP is its speed, as it uses a smaller number of proxies, the free version will give you bandwidth up to 30–50 Kbps, but in premium services, you can opt for a price of 5 euros/month. The main JAP page is presented in Fig. 17.

You can find more JonDonym anonymous proxies in config settings that work on port 4001. The JAP browser works exactly like the TOR browser. You can create a new identity. Also, the browser comes up with add-ons such as No-Script and HTTPS everywhere.

The working of JAP is shown in Fig. 18.

To make JAP more secure, you can use JAP network in the TOR browser, as TOR gives you more reliability in order to become anonymous. To do that, configure TOR proxy by putting port 4001 (JAP port)—127.0.0.1:4001 (HTTP proxy). This is very restricted to HTTP as the free version supports only HTTP. In the premium version, you will be able to use any port and any other SOCKS proxy.

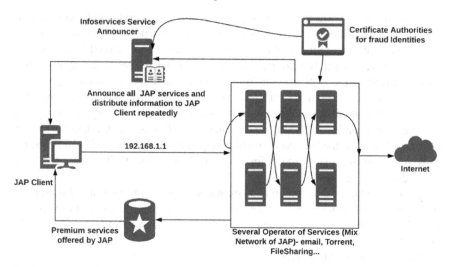

ⓘ New services found! (Click here)

🛈⑤ SpeedPartner ▦ - Cyrax ▯ ▮   ▼ ⟳   Config...

SpeedPartner-Cyrax:

   Number of users: 220 / 600

   Operators: ▦ ▦

   Speed: 100-300 kbit/s

   Response time: ≤ 750 ms

   Exit IP address: 178.33.255.188 ▯ ▮

JONDO STATUS   CONNECTED

DISTRIBUTION   USERS

Anonymity
◉ On
○ Off

Remaining credit: Pay now

Encrypted data transferred: 3,728.5 KByte Activity: ⌐⌐⌐⌐⌐⌐

Help other people (anti censorship) ☐ On   Activity: ⌐⌐⌐⌐⌐⌐

Fig. 17 Java Anon Proxy (JAP) page. The page shows all network and user configurations that can achieve proper anonymity. Currently, JAP is connected to a proxy located in Cyrax, and the page gives information about the number of users, speed, and exit IP address

Infoservices Service
Announcer

Certificate Authorities
for fraud Identities

Announce all JAP services and
distribute information to JAP
Client repeatedly

192.168.1.1

Internet

JAP Client

Premium services
offered by JAP

Several Operator of Services (Mix
Network of JAP)- email, Torrent,
FileSharing...

Fig. 18 Working of Java Anon Proxy (JAP). The client sends a request with a port 4001 for a service, which comes to a Mix Network of JAP (anonymity maintained here, as there is no straight path to the destination server), which forwards requests to the intended server

6 Summary

Anonymous networks and services are a vast field in research and development that are continuously increasing. However, there are still many projects ongoing to support and protect our privacy. The TOR and the I2P networks are the two systems that people are more likely to use in the following years and are gaining more and more popularity. Table 1 shows a comparison between TOR, I2P, and JAP based on various parameters such as port numbers, encryption standards, and anonymity level, which will help users to select the best among these. Also, the authors of [17] use machine learning classifiers to test the degree of anonymity level of different systems.

Table 1 Comparison between various anonymous networks—Tor, I2P, and JAP

TOR	I2P	JAP
Provides anonymity (your identity is completely unknown)	Provides anonymity (your identity is completely unknown)	Provides pseudonymity (you are not using real identity)
Works on the principle of "Onion Routing"	Works on the principle of peer–peer communication with Garlic Routing	Works on the mixing of the data stream to obfuscate the outsider
Comes up with secure browser—TOR browser	Comes up with a secure console panel—"I2P router", used in any default browser	Comes up with a secure browser—JonDo browser
Speed is slow as it covers a pool of relays to create a circuit, plus encryption at each layer	Speed is slow, as it also has to cover a series of peers to connect, plus encryption at each gateway	Speed is fast, as it consists of a smaller number of nodes
Works on ports 9001 and 9030 for network traffic and directory information	Works on ports 7657 and 4444 to start the I2P router and to use eepsites	Works on port 4001 to surf the Internet anonymously
Encryption—AES with Diffie–Hellman for key exchange	Tunnel Encryption—AES Garlic Encryption—Elgamal/SessionTag + AES Transport Encryption—DH/STS + AES	Encryption—RSA 1024 + AES 128 GCM (Galois/Counter Mode)
Not Decentralized	Completely decentralized	Uses a number of mixes of JAP network
Ongoing research and open source	Ongoing research and open source	Standard technology

References

1. PewResearch, *The state of privacy in America* (Pew Research Center, 2020). https://www.pew research.org/fact-tank/2016/09/21/the-state-of-privacy-in-america/. Accessed 11 Sept 2020.
2. The state of privacy in America (Pew Research Center, 2021). [Online]. Available: https:// www.pewresearch.org/fact-tank/2016/09/21/the-state-of-privacy-in-america/.
3. G. Sterling, Report: almost 90 percent concerned about online privacy & trying to avoid advertisers, in *MarTech* (2021). [Online]. Available: https://martech.org/report-almost-90-percent-concerned-about-online-privacy/.
4. L. Rainie, S. Kiesler, *Anonymity, Privacy, and Security Online* (Pew Research Center: Internet, Science & Tech., 2020). https://www.pewresearch.org/internet/2013/09/05/anonym ity-privacy-and-security-online/. Accessed 12 Sept 2020.
5. M. Cabric, Confidentiality, integrity, and availability, in *Corporate Security Management* (Butterworth-Heinemann (UK), Springer, 2015), pp 185–200.
6. TOR, *The Tor Project | Privacy & Freedom Online*, in Torproject.org. (2020). https://www.tor project.org/. Accessed 4 Nov 2020.
7. R.A. Haraty, B. Zantout, The TOR data communication system. J. Commun. Netw **16**(4), 415–420 (2014)
8. T.G. Abbott, K.J. Lai, M.R. Lieberman, E.C. Price, Browser-based attacks on tor, in *Privacy Enhancing Technologies. PET 2007*, ed. by N. Borisov, P. Golle. Lecture Notes in Computer Science, vol 4776 (Springer, Berlin, Heidelberg, 2007).
9. S.J. Murdoch, G. Danezis, Low-cost traffic analysis of Tor, in *2005 IEEE Symposium on Security and Privacy (S&P'05)* (Oakland, CA, USA, 2005), pp. 183–195.
10. E. Çalışkan, T. Minárik, A.-M. Osula, Technical and legal overview of the tor anonymity network, in *NATO Cooperative Cyber Defence Centre of Excellence*, Estonia.https://ccdcoe. org/uploads/2018/10/TOR_Anonymity_Network.pdf. Accessed 26 Oct 2020.
11. E. Cambiaso, I. Vaccari, L. Patti, M. Aiello, Darknet security: a categorization of attacks to the tor network, in *CEUR Workshop Proceedings*, vol 2315 (2019).
12. I2P, I2P Anonymous Network, in Geti2p.net (2020). https://geti2p.net/en/. Accessed 4 Aug 2020.
13. N. P. Hoang, P. Kintis, M. Antonakakis, M. Polychronakis, An empirical study of the I2P anonymity network and its censorship resistance, in *Internet Measurement Conference* (2018), pp. 379–392.
14. J. P. Timpanaro, T. Cholez, I. Chrisment, O. Festor, Evaluation of the anonymous I2P network's design choices against performance and security, in *2015 International Conference on Information Systems Security and Privacy (ICISSP)*, pp. 1–10, Angers, 2015.
15. I. Clarke, O. Sandberg, B. Wiley, T.W. Hong, Freenet: a distributed anonymous information storage and retrieval system, in *Designing Privacy Enhancing Technologies*, ed. by H. Federrath. Lecture Notes in Computer Science, vol 2009 (Springer, Berlin, Heidelberg, 2001).
16. Freenet—Wikipedia (En.wikipedia.org, 2021). [Online]. Available: https://en.wikipedia.org/ wiki/Freenet.
17. A. Montieri, D. Ciuonzo, G. Aceto and A. Pescapé, "Anonymity Services Tor, I2P, JonDonym: Classifying in the Dark (Web)" in *IEEE Transactions on Dependable and Secure Computing*, vol. 17, no. 3, pp. 662–675, 1 May-June 2020.

Chapter 3
TOR—The Onion Router

1 Introduction

We had seen a continuous advancement in privacy-related technologies for several years, which have improved and are still revamping. Privacy is a particular need after seeing several scaled surveillance by the government or its allied corporations. To protect from such surveillance, people started using anonymous software and applications to conceal their identities. In typical cases, when you are connected to the various distributed systems or Internet, you reveal IP address in any server's logs. Even if you try to connect using a proxy server or any other nefarious way to hide your identity, your Internet Service Provider (ISP) always has a keen eye on you. Furthermore, connecting to the Internet with someone's IP address (computer café, airport Wi-Fi, friend's IP address) is also not completely safe in terms of anonymity. You can still be identified via monitoring your sessions through social media or from your e-mails. It's not always government, but it can be an attacker who wants to steal the data and later can be used as a weapon for identity theft or may be used for selling the data to various other corporations. People leave a tremendous amount of private information when they are connected to the Internet, such as saved passwords, credit/debit card numbers, and photos. This information can later be fetched out with any forensic or analysis technique, like footprinting, which can reveal the user's true identity and hence lose privacy.

In 2018, the Facebook Cambridge Analytica data scandal was one of the biggest data scandals in history. Millions of personal data of Facebook users get harvested and gained in political advertising. [1] (See: "Facebook says data leak hits 87 million users, widening privacy scandal").

Another devious data breach happens at verification.io, a big data company that verifies e-mail, where around 763 million records gone public, discovered by a security expert Bob Diachenko. In his blog post, he had written—"Database of MongoDB of size around 150 GB no password-protected data has been discovered, where a massive number of e-mails linked up with phone numbers get accessible for anyone

© The Author(s), under exclusive license to Springer Nature Singapore Pte Ltd. 2022
N. Dutta et al., *Cyber Security: Issues and Current Trends*, Studies in Computational
Intelligence 995, https://doi.org/10.1007/978-981-16-6597-4_3

```
"_id" : ObjectId('█████████████████████'),
"zip" : '█████',
"visit_date" : ISODate("2017-11-28T03:30:00.000+0000"),
"phone" : '████████████',
"city" : "██████████",
"site_url" : "studentsreview.com",
"state" : "IA",
"gender" : "female",
"email" : █████████████,
"user_ip" : '████████,
"dob" : ISODate("████████████0:00.000+0000"),
"firstname" : ████████,
"lastname" : █████████,
"done" : 1.0,
"email_lower_sha256" : "███████████████████████████"
```

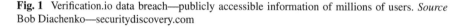

Fig. 1 Verification.io data breach—publicly accessible information of millions of users. *Source* Bob Diachenko—securitydiscovery.com

who is having an active internet connection". [2] (See: "Verification.io suffers major data breach"). The data is accessible publicly, as presented in Fig. 1.

In the same queue of data breaches, India also lacked in imposing privacy policies. Approximately 1.1 billion people data gets vulnerable because of the expanded Aadhaar biometric system, which connects every citizen in the country and is one of the largest and biggest biometric ID systems in the world, owned by the Unique Identification Authority of India (UIDAI). The system includes every citizen in the country and their fingerprints, addresses, photographs, and other personal information such as date of birth and phone numbers. With the Aadhaar ID (similar to a social security number used in US)—a 12-digit number, you can open a bank account, get a SIM card, one can vote, and use any government service that asks for identification. The citizen's data of Aadhaar ID gets breached by the authority who is intended to protect it. People sell the login credentials of Aadhaar on WhatsApp, where anyone can enter the Aadhaar number to access all the information related to the ID and can sell it for nearly $7 USD [3].

In addition to the scam mentioned above, the Marriott breach is one of the most significant data breaches with powerful tools that can grab names, payment details, e-mail, mailing addresses, passport, and credit card details which influence around 500 million users. The attackers used Remote Access Trojan (RAT) along with Mimikatz, shown in Fig. 2. It is a powerful tool to extract the username and password from the system memory [4] (See: "Marriott data breach").

Moreover, another major data breach in history is the Russian Secret Intelligence Agency Hacked. Hackers had targeted FSB—Federal Security Service of the Russian Federation, which is mainly responsible for counterintelligence, fight against crime, drug, human trafficking, and terrorism. Attacker heist data of around 7.5 terabytes from a contractor that exposed the confidential FSB projects, which include Nautilus (a project for collecting social media data), Nautilus-S (a project for deanonymizing TOR Traffic), Reward (a project that penetrates peer-to-peer networks), and Mentor (a project that monitor and analyzes e-mail communications) [5] (See at "Russia's Secret Intelligence Agency Hacked: Largest Data Breach in Its History").

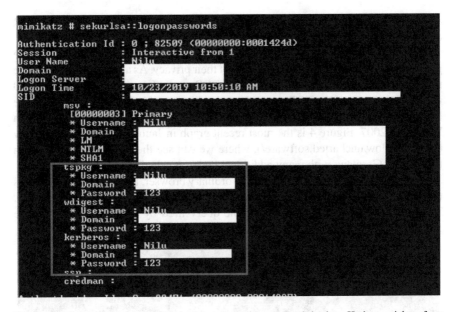

Fig. 2 Mimikatz—post exploitation tool to extract the password, hashes, Kerberos tickets from system memory

2 TOR—The Onion Router

The team of researchers and developers who formulated the idea of an anonymous network believed that Internet users must have private access to the World Wide Web. Back in the 1990s, when mass surveillance and tracking were at their peak, David Goldschlag, Mike Reed, and Paul Syverson formed the team at U.S Naval Research Lab (NRL) that thought to create an Internet connection that doesn't lose its privacy, even when someone is monitoring or sniffing the connection. They soon come up with an idea where an Internet connection or a traffic route will go through multiple servers, and at each server, the traffic will be encrypted. They call this technique "Onion Routing", further developed and maintained by Defense Advanced Research Projects Agency (DARPA).

Later, Roger Dingledine, a recent graduate from the Massachusetts Institute of Technology (MIT), joined the NRL team to work on Onion Routing. They modified the project name to TOR. Sooner, Nick Mathewson, batch-mate of Roger's, joined the team. In 2002, once the TOR network is established and conceived by many users, they put the software in the category of open source to make it more transparent and flexible. Later the code was also released under a free and open-source software license. It started with a boon of people and participated in voluntary nodes mostly from the USA and Germany. Electronic Frontier Foundation (EFF) gets fascinated by the work of Nick and Rogers and by realizing the advantages of the TOR network, EFF began funding the project in 2004. In 2006, the project was in the hand of a non-profit organization Tor Project Inc. which handles and maintains the TOR development for the betterment of user's privacy.

TOR had only a few nodes (routers) in the development stage; however, as time grows, people get to know the importance of their privacy. As a result, more and more nodes get added to the list and currently, the TOR network comprises more than 7000 nodes. The history of the TOR network can be discovered in Figs. 3 and 4, where we can see the blue nodes is the traffic going on and off from the USA to Germany in December 2007. Figure 4 is the most recent graph in January 2016, from TorFlow (https://torflow.uncharted.software/), where we can see the dense traffic flow from the USA to Germany with many additional nodes attached to the network.

When you browse the Internet with an ordinary browser, the principle it follows is the client–server architecture, where a client requests some resource from the server. The server will respond according to the query fired by the client. It is a single

Fig. 3 TorFlow [6]—simulation of TOR traffic. Blue dots show the nodes of the TOR network, which is very small in numbers between the USA and Germany in 2006

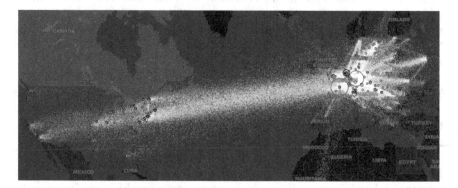

Fig. 4 TorFlow—simulation of TOR traffic. Blue dots show the node of the TOR network, which increased in numbers between the USA and Germany in 2016. These additional nodes have increased the scalability and speed of the TOR network [6]

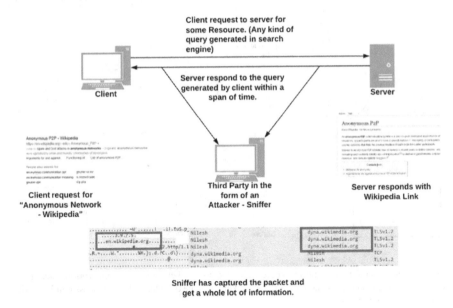

Fig. 5 Normal Internet usage—where a client requests the Wikipedia Server, in between an attacker can sniff the traffic and can analyze what information is there in the traffic

connection between the client and server, but the problem it possesses is that anyone can sniff the connection and get the idea of whom you are connecting with and what services you are using.

In Fig. 5, we can see that the client requests Wikipedia.org to search keyword "Anonymous Network" and the server responds with the same. Still, we can see the sniffer captures the packets from that single connection and gets a whole lot of information that includes DNS information, the used communication protocol, IP addresses, traffic pattern, the behavior of your search and many more that can raise serious questions to your privacy.

Now, think about a scenario where your search is on some investment, capital market, or dating website. The potential attacker gets to know your browsing interest from sniffing your traffic and leads to pushing advertisements the next time you visit the Internet.

2.1 Onion Routing

Onion Routing encrypts the message request into triple encryption, and at each node, the encryption layer is going to be decrypted and forwarded to the next node until it reaches its destination. For example, we have three nodes: Node 1, Node 2, and Node 3, and also we have a client and the dedicated server, as shown in Fig. 6.

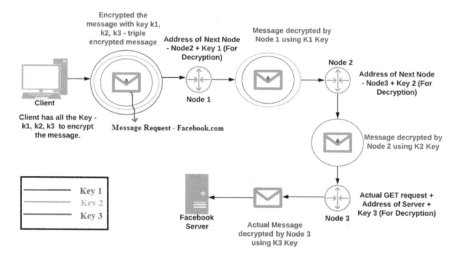

Fig. 6 TOR—The Onion Routing, where a client sends packets with triple encryption, each node will decrypt the encryption and forward the packet to the next node, and this process continues till it reaches the intended server

The client will generate a key using any key generation mechanism (elliptic curve, Diffie–Hellman). For example, let's say we generated keys as K1 (blue circle), K2 (green circle), and K3 (pink circle) that encrypts our message request "**facebook.com**" layer by layer (triple encryption—as we have three nodes).

- Triple encrypted message forwarded to Node 1, where Node 1 has only the information of the next node (address of Node 2) and the decryption key K1 to decrypt the first layer of encryption (shown with blue circle), after decrypting, the message is forwarded to Node 2.
- Node 2 has only the information of Node 3 (address of Node 3) and the decryption key K2 to decrypt the second layer of encryption (shown with a pink circle), after decrypting, the message is forwarded to Node 3.
- Node 3 is the last end of the routing, where he gets the actual GET request from the client. From Node 3 to the server, the connection is completely unencrypted (still basic Transport Layer Security (TLS) works), Node 3 has the information of actual server (in our case facebook.com). It also has the last key K3 to decrypt the last layer of encryption (shown with a green circle). After decrypting, the message is forwarded to the Facebook server.
- The Facebook server will respond to the request in the same manner but in a reverse direction with the same path and with the same nodes, where each node again started forming layered encryption using their specific key (K1, K2, and K3).

This layered encryption makes TOR one of the best options to secure and preserve your privacy, as in between, no one can know the real owner of the packet as every node just knows the next-hop address and, on top of that, the packet content is encrypted. After 2005, TOR's development is not just to develop TOR proxies, but to create a complex package, namely TOR browser, whose development started in 2008. With TOR browser, more users get started using the TOR network to access an anonymous network and preserve their privacy. Furthermore, it hides the user's identity and allows them to access an enormous amount of critical resources such as social media, hidden wikis, blocked websites, and dynamic services.

3 TOR Browser Installation

TOR browser is an extended version of Firefox, which gives anonymity while browsing the web, and it protects your identity over the Internet. You can download the browser bundle from his official website (https://www.torproject.org/dow nload/). On Microsoft Windows operating system, it will download an executable file, which you need to install. After installation, it will create two files in the installation folder- Start TOR Browser and Browser. "Start TOR Browser" is a shortcut by which you can start TOR, whereas "Browser" is an actual folder where all configuration and .dll files of TOR browser reside. When you start TOR, it will take a few minutes to establish a connection with TOR relays and fetching relay information. Once it gets completed, you will get a webpage of TOR with the default search engine Duck-DuckGo. You can check your browser's status, if it successfully installed TOR or not (you are secure or not) using the following address https://check.torproject.org/ as shown in Fig. 7. By default, the JavaScript is enabled since by disabling it, many websites will not work. However, you can still modify it in the security settings of the browser. The browser comes up with two default add-ons—HTTPS everywhere, which provides HyperText Transfer Protocol Secure (HTTPS) connection instead of HTTP, as it will block all HTTP connections and automatically use HTTPS security with a single click. Another extension is—NoScript—which enables JavaScript, Flash, Java, and other third-party plugins to be only executed by trusted sources (websites).

The other security settings that you should configure is Firefox settings > Privacy & Security. This feature can be used to manipulate certain attacks related

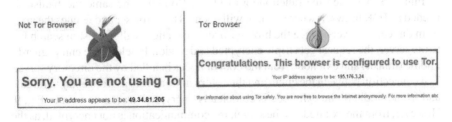

Fig. 7 TOR browser test web page—after a successful installation, it verifies that TOR is successfully configured or not. This will inform us that our identity is secured and our connection to any server is anonymous

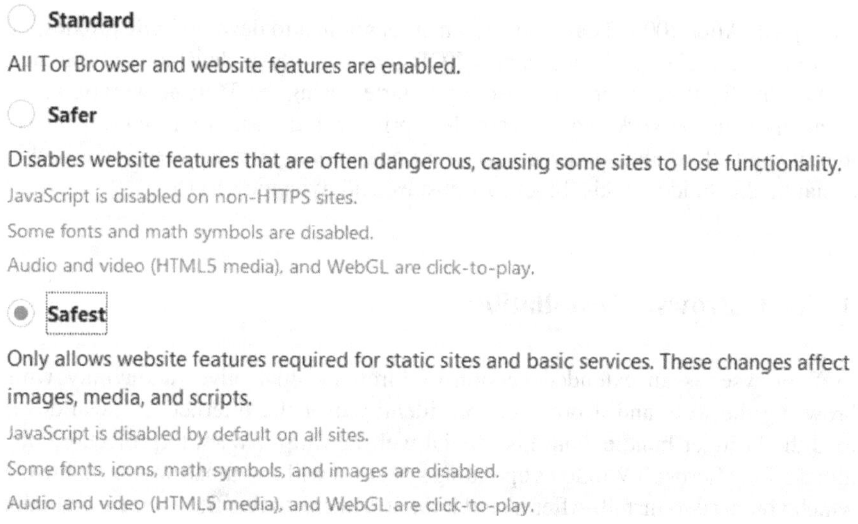

○ **Standard**

All Tor Browser and website features are enabled.

○ **Safer**

Disables website features that are often dangerous, causing some sites to lose functionality.

JavaScript is disabled on non-HTTPS sites.

Some fonts and math symbols are disabled.

Audio and video (HTML5 media), and WebGL are click-to-play.

● **Safest**

Only allows website features required for static sites and basic services. These changes affect images, media, and scripts.

JavaScript is disabled by default on all sites.

Some fonts, icons, math symbols, and images are disabled.

Audio and video (HTML5 media), and WebGL are click-to-play.

Fig. 8 TOR (Firefox) privacy & security settings with three options—standard, safer, and safest, where "Standard" is the least secure and "Safest" is the most secured connection when using TOR network

to your privacy and anonymity. The default setting is "Standard"; however, it is recommended to use "Safest", for complete anonymity, but this will surely hurt your browsing usage as many websites may not function correctly due to security settings (Fig. 8).

In addition to the above configuration, TOR will also prompt you for the configuration settings before it establishes the connection (when you start TOR browser). For example, it asks if "TOR is censored in my country", and if you select it, you can have the option to use "bridge", which is also a node. You have three options to select a bridge: either you go for the default bridge (obfs4 and meek-azure), request a bridge from torproject.org, or provide a bridge that you know from somewhere. Another setting is if you want to use a proxy to connect to the Internet, you can choose the proxy type (SOCKS4, SOCKS5, HTTPS), with its address, port, username, and password.

Figure 6 explained the functioning of Onion Routing. The same mechanism is needed in TOR browser, where a client will open a TOR browser and fetches the node from directory servers. Once the browser opens, the client will request in search bar to the server, the request gets triple encrypted and randomly selects the entry (guard), middle (transit) and exit node from the pool of nodes handled by the directory server. The connection between the client and then entry node as well as from entry to middle and from middle to the exit node is completely encrypted, as can be seen in Fig. 9. However, from the exit node to the server, the communication is not encrypted, as the request has to find the actual server to get the response and it needs to be unencrypted.

The entry, middle, and exit nodes form the TOR circuit, which can be seen in the browser as well shown in Fig. 10.

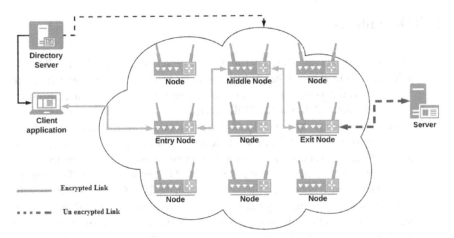

Fig. 9 Encryption flow of TOR network from client to server. The solid green line indicates complete encryption from node to node, whereas a red dashed line is an un-encrypted connection from exit node to server

Fig. 10 TOR circuit—shows the guard, middle, and exit nodes. Every circuit has a unique middle and exit node. However, the guard node will remain the same for the user

We can see in Fig. 10 that the client is requesting *duckduckgo.com,* and the circuit is made up of three nodes: The entry node is from Germany, the middle node is from France, and the exit node is from the USA, and at last, there is the requested server (duckduckgo.com).

TOR also allows you to change the circuit by pressing the blue button just below the TOR circuit. In addition to that, we can also create a whole new identity with a new circuit in between your search. You can also observe that the entry node is not changing even if you close and open the browser. Certainly, it will change if you change your network (different ISP) [7]. This challenge is refined by the team of Tor Project, who designed Changing of the Guards (COGS), where selection and rotation algorithms were used when selecting the entry guard.

4 TOR Entities

(1)   TOR directory server—It maintains several servers that have a list of all available active TOR relays. It helps the client in building the TOR circuit whenever a user opens the TOR browser. It is the first and foremost entity a TOR client gets to interact with to use the TOR services.

(2)   TOR client—It can be specialized hardware or software which runs on the computer to access the TOR network. It simply asks the service from the server using a GET request and the server will respond according to the service asked for. In our case, the person who installed the TOR browser is our TOR client.

(3)   TOR entry (guard) node is the first relay node in the TOR Circuit, which directly interacts with the client. Any client who wants to connect to the TOR network has to go through the entry node. It only knows the IP address of the next node (middle node) and a decryption key to decrypt the first layer of encryption (message is triple encrypted). The IP address of a guard node will never change, even if you close and end the session of TOR browser, until and unless you change your network. However, the IP addresses of other nodes get changed. Few websites, like dan.me.uk/, torstatus.blutmagie.de/, check.torproject.org/, list the currently available guard nodes. The guard node usually changes its IP address approximately every 2–3 months.

(4)   TOR middle (Transit) node—It is the second and most prominent node in the TOR Circuit, which maintains the user's anonymity. The transit node only knows the IP address of the next node (exit node) and the decryption key to decrypt the second layer of encryption. This node doesn't know where the network traffic is going to where it is coming from, except it only knows to forward the traffic to the next known node.

(5)   TOR exit node—This node accepts the actual GET request from the client. However, it doesn't know who the real client is, as it only knows where to forward the packet—to the "intended server"—and the decryption key to decrypt the last layer of encryption. It forwards the decapsulated packet to the server in a completely unencrypted format. The IP address of the exit node is directly visible to the intended server and any intruder easily fetches it by doing simple sniffing into the network. So, this node is the most affected in the TOR Circuit, where many malicious attempts occur to deanonymize the client and its traffic pattern. The TOR exit nodes list are available as well on various websites like dan.me.uk/, torstatus.blutmagie.de/, check.torproject.org/.

(6)   Bridge—Sometimes, direct access to the TOR network is blocked by the ISP or by the government. In such cases, an alternative node that can access the TOR network is required. Such nodes are called "pluggable transport" or simply bridges. They are the unpublished nodes not listed in the public directory of TOR nodes. However, there are repositories where the entry and exit nodes lists are already public that anyone can see, and therefore the nodes can be blocked by ISP. Such cases happened in China, where TOR is completely restricted, as all the traffic is monitored by a mechanism of traffic analysis known as "deep

packet inspection (DPI)". They analyze the packet to its root and if found any IP associated with any TOR node (from the lists of publically available nodes), it simply blocked it. The authors of [8] revealed how China blocks TOR traffic. Alice (a TOR user) cannot access the TOR website as it is blocked by the Chinese government; however, there are mirror website where one can access TOR. Let's assume that somehow Alice is able to download TOR software bundle, but when he starts TOR network, TOR will not be able to connect to the directory authorities to fetch public TOR relays as it again blocked on the IP layer. More than bypassing the restrictions in China, bridges can be very helpful as they are not listed anywhere and no one knows where precisely the node is located. So, instead, the connection is first going to the entry node, now it—goes to the bridge node, and then it moves to the entry node. Though they are not secure as it is unpublished and no one knows who maintains them, they can be the most eminent point from an attacker's perspective, where an exploit can be plugged.

There are three simple ways you can get the bridge node. The first is by using the TOR settings, where you can select the default bridge (obfs4 and meek-azure). The second option is you can request the bridge from torproject.org, and they will provide you three bridges, as shown in Fig. 11.

The third option to get the bridge is to obtain it from some reliable source and add it to the space given. An alternative way to get the bridge is you can simply mail to bridges@torproject.org using either Gmail or Riseup, and they will provide you three bridges in the mail itself. Generally, your bridge will look like this:

obfs4 87.162.112.114:8080 59EB1E1FB8658DA569BEB012C26D695D320C93 60cert=Becy7LHoHCApnHZsY0ZCbLL7AJM0JGTHUMUw0W1hVDMQULu03w/ 0B3CjEksuRKyzyZTvZQ iat-mode=0.

Here, obfs4—is the pluggable transport technique used in the bridge, 87.162.112.114—is the IP address of the bridge node, 8080—port number, and the leftover is cert—which is the unique identifier of a bridge, iat-mode—is the inter-arrival mode (time between each arrival into the system and the next), it is used to conquer the dpi fingerprints that based on timing so that dpi don't block the traffic.

⦿ Request a bridge from torproject.org   Request a New Bridge...

```
obfs4 87.162.112.114:8080 59EB1E1FB8658DA569BEB012C26D695D320C9360 cert=Becy.
obfs4 94.177.160.181:110 D2DFC36FCCAFA723F0756BEE8C33B45B46D1A41C cert=dtGU/
obfs4 186.203.230.72:37277 CE27CF5CF242149830130140F13660A479FDBA60 cert=D4w1
```

Fig. 11 Pluggable transport—TOR bridge, these are the unpublished nodes that are not listed in the public directory of TOR

5 TOR Status

After TOR is configured, we visualize and analyze various system and network parameters like bandwidth, CPU usage, memory usage, upload, and download speed by using *nyx,* which is a terminal-based monitoring tool for the TOR network. It is written in Python and available under GNU General Public License. *nyx* provides real-time statistics and information on the TOR network. The information includes resource usage (bandwidth, CPU, memory), relay (node) information, and log information with regex filter and torrc configuration. It runs on every UNIX-based platform (Mac, Linux, OSX, and BSD) but not on the Microsoft Windows operating system.

On Linux systems, just start the Terminal and type the command *nyx*. Then start your TOR browser and the *nyx* tool will automatically start providing the network statistics, as shown in Fig. 12.

Nyx will also prepare an event log, such as the one presented in Fig. 13, where it is showing you the opening and closing of the TOR browser with its specific time. This piece of information could be essential to a forensic investigator working with a TOR network.

Furthermore, there is a huge bunch of information in the log file that you can read and analyze for further investigation on the TOR network. For example, we queried for the "hidden wiki" and we got this URL—http://hiddenwikitor.com/. Therefore, when you check your log status, you will find this URL and the IP addresses with its

cpu: 0.0% tor, 1.1% nyx mem: 14 MB (1.5%) pid: 5661 uptime: 03:15
page 1 / 5 - m: menu, p: pause, h: page help, q: quit
Bandwidth (limit: 1 GB/s, burst: 1 GB/s):
Download (0.0 B/sec):
Upload (0.0 B/sec):
2 KB
1 KB
543 B
0 B
543 B
362 B
181 B
0 B
5s 10 15 20 25
5s 10 15 20 25

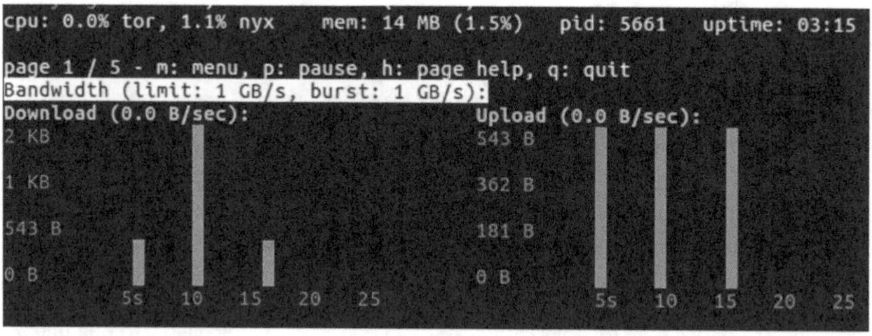

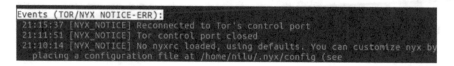

Fig. 12 Nyx—a visualization tool for TOR network. A green bar shows the download maximum speed/sec, and the blue bar shows the upload maximum speed/sec. CPU and bandwidth usage are also visible

Events (TOR/NYX NOTICE-ERR):
21:15:37 [NYX_NOTICE] Reconnected to Tor's control port
21:11:51 [NYX_NOTICE] Tor control port closed
21:10:14 [NYX_NOTICE] No nyxrc loaded, using defaults. You can customize nyx by
placing a configuration file at /home/nilu/.nyx/config (see

Fig. 13 Nyx event viewer—it will show the events related to the TOR browser, such as—opening and closing TOR ports and errors

21:21:49 [STREAM] 22 NEW 0 hiddenwikitor.com:80 SOURCE_ADDR=127.0.0.1:56672 PURPOSE=USER
21:21:49 [CIRC_BW] ID=17 READ=1018 WRITTEN=1018 TIME=2019-11-02T04:21:49.235771 DELIVERED_R
  OVERHEAD_WRITTEN=758
21:21:49 [CIRC_BW] ID=15 READ=9162 WRITTEN=509 TIME=2019-11-02T04:21:49.235758 DELIVERED_RE
  OVERHEAD_WRITTEN=108
21:21:49 [STREAM_BW] 21 390 8518 2019-11-02T04:21:49.235735
21:21:48 [CIRC] 17 BUILT
  $D27208881BBDB5EA56EFD1D1799187519591E325~niftytellcomys,$C156F1DD69EEFC061CFC34F19E1C561
21:21:48 [BUILDTIMEOUT_SET] COMPUTED TOTAL_TIMES=286 TIMEOUT_MS=1890 XM=1075 ALPHA=2.850166
  CLOSE_MS=60000 CLOSE_RATE=0.000000

Fig. 14 Nyx log viewer—it will log the details of the TOR connection to which you are connecting. Here hiddenwikitor.com is a web page with its port and IP address logged in nyx

Fig. 15 Nyx logs the complete circuit of TOR in its log file, revealing the relay nickname—"KagamineLenTwilight" with its Onion Routing fingerprint

port number and various other relevant data that can be used for forensics purposes, as presented in Figs. 14 and 15.

Other details can be fetched in *nyx* by pressing "m" and selecting the suitable option. With that, if you select the option "connection", you will get the TOR Circuit in the terminal, and if this circuit is logged somewhere, then it is the most important asset from the forensic perspective that we will see in our later chapters.

Additionally, you can also visit these websites to know more about your real-time TOR connection and how you can improve it—https://metrics.torproject.org/services.html, https://metrics.torproject.org/rs.html, https://torstatus.rueckgr.at/.

6 TOR for Mobile—Orbot

The number of mobile devices is increasing along with their processing power and hardware resources. There are 5 billion people using smartphones daily as they are cheap and portable compared to the desktop and laptops. Hence, we need to preserve our anonymity on such platforms too. Therefore, TOR comes up with a new open-source application, "Orbot" for smartphones. It acts as a proxy in smartphones that allows other applications to use the Internet more securely. It works similar to TOR as it encrypts the traffic between the client and server using the same Onion Routing mechanism and jumping the request from one node to another in an encrypted form. The intention of TOR and Orbot is the same: to hide your true identity against mass surveillance and to protect your personal freedom and privacy.

You can get the Orbot application from the Google Play Store. After installing it, you will get the interface presented in Fig. 16. You can start the Orbot proxy by pressing the onion icon (grayed initially, but after starting, it becomes yellow). It will identify TOR nodes and in a few seconds, it will connect to the TOR network. After connecting, it will show both the download and upload speeds.

You can configure the proxy the way you want, such as you can connect to any specific country by selecting the country from the option "Global (Auto)". Also, you can use bridges to get more secured, as described in Sect. 4 of this chapter. Orbot contains HTTP and SOCKS4A/SOCKS proxies with port numbers 8118 and 9050. It can transfer all the TCP traffic from the application which uses the normal Internet into TOR traffic. The logging information can be fetched by swiping right, as shown in Fig. 17.

An alternative to Orbot is "TOR browser" for smartphones. The difference between these two applications is that Orbot transfers all application (installed mobile application) network traffic via its SOCKS proxy. In contrast, TOR browser will only provide you the flexibility of using the Internet on a browser (TOR browser) that is

Fig. 16 Orbot—mobile application comes up with TOR functionality with VPN mode, where you can choose a specific application to turn its traffic in an encrypted tunnel mode

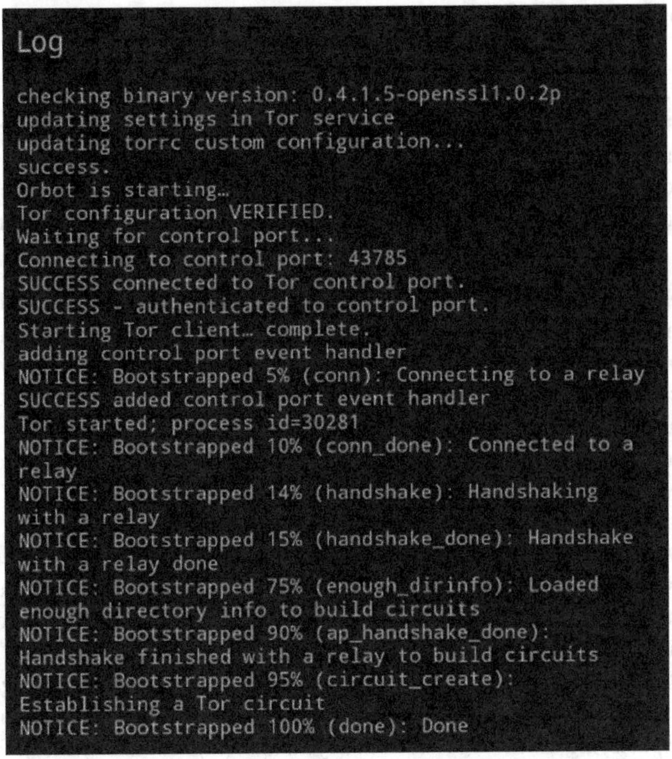

Fig. 17 Orbot events logs—shows the TOR version numbers, services, port numbers, TOR Circuit information, and status of connection establishment to relays

specifically meant for smartphone. Once you connect to the browser, it will start logging relay information in the log, and once it is connected, the browser will start as a normal browser, with a default search engine as DuckDuckGo. In the log files shown in Fig. 18, you can see the TOR Circuit with its name and fingerprints, although it will not show the circuit as we had seen in the TOR browser for desktop.

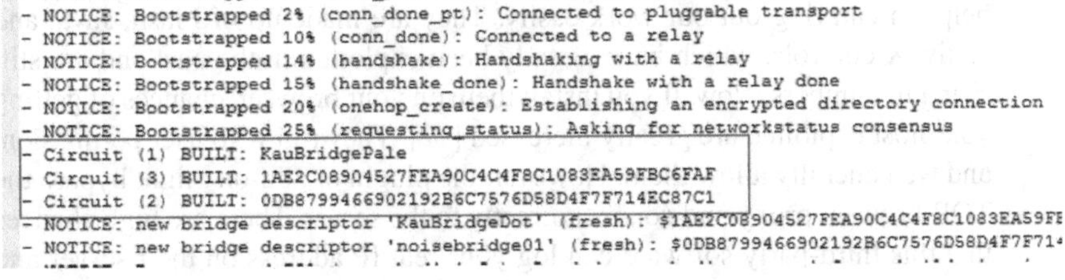

Fig. 18 Event logs—TOR browser for mobile. It shows the status of connection establishment to the TOR relays

7 Loopholes in TOR

Certainly, the TOR system is not perfect. There are many benefits, but there are also loopholes in the TOR system that can deanonymize and reveal the services you are using behind those anonymous systems. Moreover, as we had seen that the TOR exit node is completely unsafe (no encryption), it might be the most suitable entry point for the attacker to monitor the traffic and find the pattern inside it that can reveal the identity.

(1) TOR Guard Selection—As the entry node can be sniffed using simple sniffers (tshark, Wireshark), it is relevantly easy for an intruder to persuade a malicious guard node, handled by the attacker and in its ownership. He can now see the packets going in and out of the network that can assist in end-to-end correlation attacks [9]. The same attack is applied for the exit node, which is more powerful compare to the guard node, as the exit node is directly connected to the real server and the attacker can precisely know about the GET and POST request from the client.

(2) Eavesdropping attack—Autonomous systems (AS) sit at the very end of the network. If the AS sits on both sides as a client (guard node) and as well on the server-side (exit node), then it might be possible to have a correlation attack by comparing the entry and exit node traffic. In reference [10], the authors develop LASTor that can deliver latency gain while choosing the path, the path is tunable and the user can tradeoff between latency and anonymity by specifying value 0 (lowest latency) and 1 (highest anonymity). They also develop an algorithm that identifies the AS, which can correlate the traffic and ignores such AS for path selection. The other eavesdropping attack is on the exit node [11]. It seems it is an easy option as TOR doesn't encrypt the traffic between the exit node and the targeted server (although you might use HTTPS), and anyone can capture the traffic passing through it. Though it is not an easy task, it will surely not reveal the identity of the source client as the exit node has only the information about the middle node and not the client. However, the third party, which sometimes carries the traffic from exit node to server, can expose the data packets and the information.

(3) Plugin or Add-on-Based attack—Plugins are third-party software and generally, we don't trust them. Usually, the add-ons are browser-based software that helps in carrying out our work easily. They are made up of Flash, Java, and ActiveX controls, which have already been exploited in the past and are still rising in numbers. Now, if you install them on your browser, chances of getting your host exploited are greatly increased [12]. They run with user permission, and we generally allow them. However, the plugin or add-on might bypass the TOR browser proxy and connect directly to the server. There are high chances that this third-party software can log your real IP address on their server and again disclose the true identity.

(4) Bad Apple attack—In March 2011, French Institute for Research in Computer Science and Automation had documented about Bad Apple attack [13], which

divulge the real IP address of BitTorrent users who are using torrent on the TOR network. Therefore, it is strictly recommended on TOR documentation not to use any third-party software which can log the real IP addresses.

(5)   Sniper attack—this is a low-cost but strongly destructive denial-of-service (DoS) attack that uses valid protocol messages to consume memory by exploiting TOR end-to-end reliable data transport [14]. It also enables the deanonymization of various hidden services by using selective DoS that choose only the guard nodes that control the adversary.

(6)   Heartbleed bug—It is the vulnerability in the OpenSSL library. The vulnerability allows looting the information that is protected. With this bug, anyone on the Internet can read the system's memory protected by OpenSSL software. The anonymity of a client in TOR can be compromised if an intruder can grab the encrypted information from the node. Therefore, to avoid this notion, Roger Dingledine identified and rejected almost 380 vulnerable exit nodes running OpenSSL and blacklisted from the network.

Other attacks also might hinder in restricting anonymity, such as timing attack and correlation attack [15, 16]. A simple Wireshark sniff is not enough to disclose the TOR Circuit, as Wireshark will only fetch the guard IP address and not the further node IP address, as shown in Fig. 19. However, the sniffer can tell that you are using TOR, but it can't disclose what you are doing in TOR.

Though there are several mechanisms we can use to deanonymize the identity, deep packet inspection (DPI) [17] is one such mechanism that can sniff each packet and its label moving from router to router and determines where to send it. Firewalls use this packet label to identify the traffic and the governments are more concerned about using such firewall boxes equipped with DPI. Some of the best DPI are— nDPI (https://www.ntop.org/products/deep-packet-inspection/ndpi/), bro (www.zeek.org), or netify (https://www.netify.ai/developer/netify-agent).

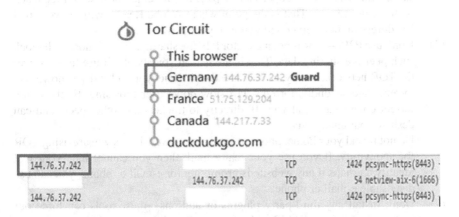

Fig. 19 TOR circuit shows the guard node. Sniffers such as Wireshark can easily capture the guard node in the network traffic

8   What not to Use in TOR

Solidly, TOR is the best tool to preserve your identity and privacy, but you have to install and use it with precise care. There are numerous warnings while using the TOR network, and some of them are summarized below.

(1)   Don't use Windows! The Microsoft Windows operating system is best for business and educational purposes, but it is never suited to preserve anonymity. With the latest version of Windows 10, we can see that the system is completely cloud-based, and the cloud is easily negotiable from a security perspective. Besides this, there are already countless vulnerabilities and security bugs in Windows that it never going to stand in conserving user's privacy. Perhaps, you can rely on Linux operating system or TOR configured Linux distributions like Tails (www.tails.boum.org) or Whonix (www.whonix.org).

(2)   Don't use HTTP websites! Although TOR is perfectly secure and encrypted, never forget that after the exit node to the server, the connection is completely insecure and it is the most favorable target point for any intruder. So, in case if you are using an HTTP website, the traffic itself will divulge the data inside the packets. Instead, you should use "HTTPS everywhere" as it is the default add-on in TOR browser that will automatically switch HTTP traffic to HTTPS for the supported website.

(3)   Don't use JavaScript, Java and Flash!—These extensions are pretty nasty as they run with user account's privileges, possibly can access the data. Java and Flash run in virtual systems, and they ignore the TOR proxy settings; hence, they bypass the TOR security protection.

(4)   Never download from TOR browser! TOR network has vast anonymous hidden services intended to do nefarious activity with you or your system. For example, if you download the Adobe Reader from TOR and install it in our system after successful installation, it opens a web page in our default browser for feedback or any other reason. That is the point where Adobe Reader will log your real IP address in their server and you lose your privacy [18].

(5)   Don't use P2P—TOR is not meant for P2P file sharing, and exit node will block such peer file sharing data. That is the reason you shouldn't use torrents over the TOR network. Firstly, you violate the TOR network rules if you download torrents and secondly, torrents will slow down user browsing. Furthermore, torrent clients can send your IP directly to trackers and other peers and can destroy your anonymity.

(6)   Do not reveal your home, phone, e-mail on any website, as you are using TOR for anonymity. If you use your real e-mail, then you cannot hide your real identity. Perhaps if any website is prompting for e-mail or phone, you can use disposable e-mail or phone.

(7)   Don't install any third-party plugins or add-ons—these tools can force our system to reveal our real IP address, as such tools run on a virtual system and can bypass the TOR protection. Not only this, but these tools can have other

vulnerabilities and bugs that can exploit the system or can install malware which again hinders our anonymity.

References

1. D. Ingram, Facebook says data leak hits 87 million users, widening privacy scandal. In: U.S. (2021). https://www.reuters.com/article/us-facebook-privacy/facebook-says-data-leak-hits-87-million-users-widening-privacy-scandal-idUSKCN1HB2CM
2. P. Report, (2021) Verification.io suffers major data breach. In: GRC World Forums. https://www.grcworldforums.com/breaches-and-vulnerabilities/verificationio-suffers-major-data-breach/85.article
3. G. Mengle, Major aadhaar data leak plugged: French security researcher. In: The Hindu. (2021). https://www.thehindu.com/sci-tech/technology/major-aadhaar-data-leak-plugged-french-security-researcher/article26584981.ece
4. J. Fruhlinger, Marriott data breach FAQ: how did it happen and what was the impact?. In: CSO Online. (2021). https://www.csoonline.com/article/3441220/marriott-data-breach-faq-how-did-it-happen-and-what-was-the-impact.html
5. Z. Doffman, Russia's secret intelligence agency hacked: 'Largest data breach in its history'. In: Forbes. (2021). https://www.forbes.com/sites/zakdoffman/2019/07/20/russian-intelligence-has-been-hacked-with-social-media-and-tor-projects-exposed/#6f9523256b11)
6. "Torflow", Torflow.uncharted.software, (2021). [Online]. Available: https://torflow.uncharted.software/
7. T. Elahi, K. Bauer, M. AlSabah, R. Dingledine, I. Goldberg, "Changing of the guards: A framework for understanding and improving entry guard selection in Tor", Proceeding ACM Workshop Privacy Electronic Society (WPES'12), pp. 43–54, (Oct. 2012)
8. P. Winter, S. Lindskog, How China is blocking Tor. arXiv 1-21. (2012)
9. M. Imani, A. Barton, M. Wright, Guard sets in Tor using as relationships. Proc. Priv. Enhancing Technol. (1), 145–165 (2018)
10. M. Akhoondi, C. Yu, H. Madhyastha, LASTor: a low-latency as-aware Tor client. IEEE Symp. Secur. Priv, San Francisco, CA, USA **20–23**, 476–490 (2012)
11. K. Zetter (2021) Rogue nodes turn Tor anonymizer into eavesdropper's paradise. In: Wired. https://bit.ly/2P1MDCN. Accessed 17 Mar 2021
12. T. Abbott, K. Lai, M. Lieberman, E. Price, *Browser-Based Attacks on Tor, Privacy Enhancing Technologies* (Springer, Ottawa, Canada, 2007), pp. 184–199
13. S. Le-Blond, P. Manils, C. Abdelberi, M.A. Kâafar, C. Castelluccia, A. Legout, W. Dabbous, "One bad apple spoils the bunch: exploiting P2P applications to trace and profile tor users". CoRR, (2011)
14. R. Jansen, F. Tschorsch, A. Johnson, B. Scheuermann, "The sniper attack: anonymously deanonymizing and disabling the Tor network". Proceeding 21st Annual Symposium NDSS. (California, USA, Feb 2014), pp. 1–15
15. J. Salo, "Recent attacks on Tor". In: Aalto University, Finland, (2010)
16. G. He, M. Yang, X. Gu, J. Luo, Y. Ma, "A novel active website fingerprinting attack against Tor anonymous system". Proceedings of the 2014 IEEE 18th International Conference on Computer Supported Cooperative Work in Design (CSCWD), 2014. (Hsinchu, Taiwan, 21–23 May 2014), pp. 112–117
17. F. Saputra, I. Nadhori, B. Barry, "Detecting and blocking onion router traffic using deep packet inspection", 2016 International Electronics Symposium (IES). (Denpasar, Indonesia, 29–30 Sept 2016), pp. 283–288
18. P. Chaabane, Manils, M. Kaafar, "Digging into anonymous traffic: a deep analysis of the Tor anonymizing network". Fourth International Conference on Network and System Security. (Melbourne, VIC, Australia, 1–3 Sept. 2010), pp. 167–174

Chapter 4
DarkNet and Hidden Services

1 Introduction

There is a revolution of information over the Internet that led us to the world which is getting more and more versatile in terms of technology. People read, browse, buffer, and locate information from the Internet, and now it is nearly impossible for anyone to imagine computing without access to this worldwide network. Also, the Internet is the backbone by which everyone is connected anytime and anywhere and the infrastructure that allowed many users to work from home during the COVID-19 pandemic.

The Internet is made up of various devices connected, and its scope is to share information from one end of the world to the other. Thus, it is nearly similar to a distributed system; many such distributed systems fasten with each other make an internet.

The world population is growing and so is the quantity of data. Countries like India and China have a tremendous amount of population, which generates millions of data, which flows over the network. The Internet population of China and India is estimated to be 772 million and 462 million, respectively. From estimating the numbers of Internet users, we can understand that it is the house of information. It is divided into three main categories: surface web, Deep Web, and DarkNet, presented in the following paragraphs.

Surface Web

The Internet we regularly use for our routine activities such as social media, e-mail, surfing, and reading from the various sites is known as the surface web. It constitutes a very small fraction of the percentage of approximately 4% of all Internet resources. Generally, your regular searches on the search engines (Google, Yahoo) are on the surface or "visible Internet". To date, search engines have indexed at least 6.35 billion web pages (https://www.worldwidewebsize.com/) and the number is still rising. The numbers are huge, but it participates only in a small portion of the Internet, as crawlers/spiders read the data, follow the links, and index the website. But many

N. Dutta et al., *Cyber Security: Issues and Current Trends*, Studies in Computational Intelligence 995, https://doi.org/10.1007/978-981-16-6597-4_4

websites are not followed by spiders and which are never indexed, and hence, it is not referred anywhere on the surface web or by any classical search engine.

Deep Web

Deep Web is the largest part of the Internet which is invisible, hidden, and deep inside the root of the World Wide Web (WWW), whose context is not indexed by normal search engines. The services included in Deep Web are hidden behind HTTP forms and usually secured by some payment module, where you can access these services by paying some fees to the service provider. The content can be easily accessible through any normal search engine, but certainly, you have to log in, fill up and complete all the other attributes needed in identifying you, and then you can search for a specific thread from the service database. Possibly, it is not searched by any search engines as the website is not indexed, so you have to hardcode the URL and paste it on the URL bar. The Deep Web has numerous amenities such as on-demand video, private forums, Intranet pages, research papers, government official documents, medical records, and many more.

DarkNet

DarkNet is a hidden as well as an anonymous network. DarkNet services are encrypted and no one knows their actual source. It constitutes a smaller portion of the Internet, but the services and its content are highly evil in nature, which can put your system in real danger. Any source you find, any web link you click, any file you download from such network are not trustworthy, and there are severe consequences to use such networks and their services. DarkNet or Dark Web is not accessible using normal search engines. You need specialized software and tools to access it, such as TOR browser, Anonymous Operating System (Whonix, Tails), Invisible Internet Project—I2P, Freenet, ZeroNet, GNUNet. Normal search engines do not index Dark Web services as these services are hosted in an encrypted network to preserve the privacy of both parties (one who hosted the services and one who is going to use it). Along with inoffensive content, you can find various illegal services on DarkNet, such as drug trafficking, human trafficking, private communications, identity forging, and many more.

2   TOR and Its Hidden Service

TOR is a freely available network, which is intended to give privacy to the users against government mass surveillance. It is a completely anonymous and encrypted network where clients can host their services, possibly hidden from the surface web. There is no information you can fetch of an actual host of the hidden service. These are the services that will allow you not to reveal your actual host identity or IP address.

To connect to such services, you need to use TOR or any other anonymous network such as Freenet or I2P, as described in the first chapter. There are two perspectives while using hidden services. Either you use it for any illicit activity or use it for good ends. Criminals mostly use it, but certain portions of the DarkNet are used by good people such as human rights activists, law enforcement agencies, whistle-blowers, certain government agencies, and media agencies. Any service you host on the normal Internet is visible and can be used by everyone without any restriction. However, the same service can be published in TOR as a "TOR hidden service", and it won't be listed in conventional search engines. You can find more information about hidden/onion services from the TOR Project documentation, which is available online at https://www.torproject.org/docs/onion-services.html.en and https://blog.tor project.org/nine-questions-about-hidden-services.

In order to be able to use hidden services, we need to bypass the restrictions implemented by the network owner, being it a university network or a company IT system. For example, the network administrator might block social media services or limit access to these services or to commercial websites or entertainment ones. To tackle such obstacles in an organization, the users can generally use proxy servers, a virtual private network (VPN) or an anonymous network—TOR network in special, as presented in detail in Chap. 2. We will present herein the tactics the user uses to make his connection private and secure from intruders.

A proxy server is a normal server that acts as a gateway between the client and the Internet. If you are using a proxy, your network traffic will flow through the proxy server first and then it goes to the proxy Internet Service Provider (ISP), followed by the requested server. It changes the client IP address, so adversaries never know where the request is generated. However, this is a myth now, as proxies do not help hide the identity. It also encrypts the data, so that no one can read the data in transit. Any organization that is running strict firewall rules can easily block proxy traffic, as the IP addresses of these proxies are blocklisted in the firewall rules and whenever the traffic monitor sees the matching IP address, it is going to block it. In addition, proxy traffic can be blocked using a simple PHP script in function .php installed in an enterprise web application firewall shown in Fig. 1.

In general, every organization has an Information and Communication Technology (ICT) Department that deals with the computers network and its security. They usually deploy an Intranet so that employees can use it, and for that, each employee will register themselves on the intranet server. Once registered, ICT has IP addresses of every employee, and it is now easy for ICT to eavesdrop in any of

```
function shapeSpace_block_proxy_visits() {
    if (!is_user_logged_in()) {
        if (@fsockopen($_SERVER['REMOTE_ADDR'], 80, $errstr, $errno, 1)) {
            die('Blocked due to Proxy Access');
        }
    }
}
add_action('after_setup_theme', 'shapeSpace_block_proxy_visits');
```

Fig. 1 Block proxy server traffic using web application firewall. This PHP code can be placed inside function .php (WordPress site) and will block all incoming proxy server traffic

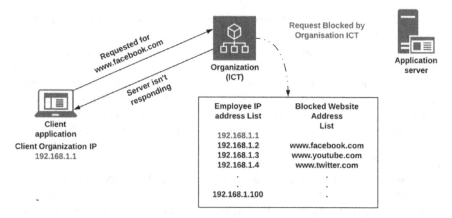

Fig. 2 An analogy of blocking normal traffic using a list where the network administrator will block websites of concern

the employee traffic. Apart from the IP address list, they also have a block website list for the employees to block certain websites such as social media, entertainment, and games, as shown in Fig. 2.

A client wants to use *facebook.com*, but the ICT Department will block this request. As described before, an access control list consists of "Employee IP address List" and "Blocked Website Address List", as presented in Fig. 2. When the ICT server processes the request, it checks with the blocked website list and the employee IP addresses. Therefore, it is easy to find who generated this request by simply using packet inspection (source, destination IP address, port number, protocol, etc.).

In another scenario, if the employee uses a proxy server, the proxy will change the source IP address to some random address. This time when the client requests the facebook.com website, the ICT will not find the IP address in the "Employee IP address List" and will treat the request as normal traffic, so it will redirect the request to the Facebook server, as shown in Fig. 3.

Virtual private network (VPN) aims to secure the privacy of the network using strong encryption policies and other security measures. It allows remote connectivity where an employee can access the private resources and corporate application using VPN from any location in the world/Internet. It works similar to a proxy but with a minute difference, such as proxy act as a man-in-the-middle server for an application such as torrent client or web browser. In contrast, VPN captures all the traffic from every application which is running on the computer and tunnels it through an encryption mechanism to tackle privacy concerns.

Anonymous networks are networks that greatly deal with users' anonymity and privacy. Such networks are peer-to-peer distributed systems where nodes are used to share information and resources anonymously with other nodes. They use a special routing mechanism (such as onion routing in the case of the TOR network) that hides the user's identity and the physical location of nodes from other nodes and networks. There are numerous such networks—I2P, Freenet, Anonymous P2P, and

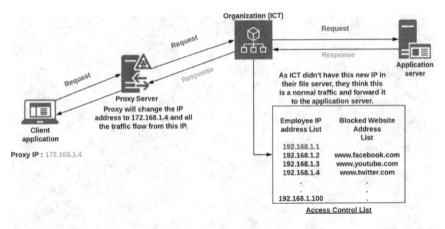

Fig. 3 An analogy of bypassing the blocked website using proxy servers

TOR. Among all of the mentioned networks, TOR is highly popularized. It consists of more than 7000 hidden nodes, over 60,000 .onion services, and around millions of users, making it the perfect and most suitable platform to hide the individual's identity.

3   Essential Concepts of TOR Hidden Services

TOR hidden services use different URL addresses compared to regular websites. Usually, in the surface web, the top-level domain (TLDs) we are using are .com, .net .org, or country-specific domains (.in, .de, .ro, etc.). The whole TOR network resembles only one TLD *.onion*, and every website should have a *.onion* extension to be called a TOR hidden service.

There is no way to connect to .onion services without using the TOR browser because of these naming conventions. It turns out that the naming convention is a strong barrier in preserving the actual host privacy since by just looking at the .onion URL, we cannot tell what service the specific website is providing, as the URL is just 16 digits random character followed by .onion extension as shown in Table 1. More than that, there is no central authority that registers and look after the .onion addresses, and hence, the TOR services are interesting pathways where hacktivism groups, law enforcement corporation, journalists, and criminal enterprises can flood their content, sometimes with an intention to harm other users in the network.

Table 1 Top-level domain and onion address		
www.facebook.com	Regular top-level domain
www.facebookcorein.onion	TOR top-level domain
http://3g2upl4pq6kufc4m.onion/	DuckDuckGo search engine

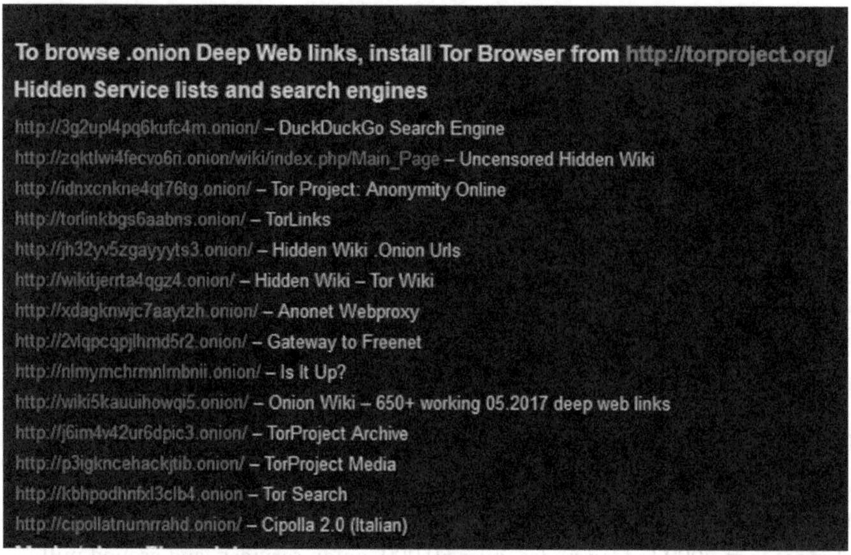

To browse .onion Deep Web links, install Tor Browser from http://torproject.org/
Hidden Service lists and search engines

http://3g2upl4pq6kufc4m.onion/ – DuckDuckGo Search Engine
http://zqktlwi4fecvo6ri.onion/wiki/index.php/Main_Page – Uncensored Hidden Wiki
http://idnxcnkne4qt76tg.onion/ – Tor Project: Anonymity Online
http://torlinkbgs6aabns.onion/ – TorLinks
http://jh32yv5zgayyyts3.onion/ – Hidden Wiki .Onion Urls
http://wikitjerrta4qgz4.onion/ – Hidden Wiki – Tor Wiki
http://xdagknwjc7aaytzh.onion/ – Anonet Webproxy
http://2vlqpcqpjlhmd5r2.onion/ – Gateway to Freenet
http://nlmymchrmnlmbnii.onion/ – Is It Up?
http://wiki5kauuihowqi5.onion/ – Onion Wiki – 650+ working 05.2017 deep web links
http://j6im4v42ur6dpic3.onion/ – TorProject Archive
http://p3igkncehackjtib.onion/ – TorProject Media
http://kbhpodhnfxl3clb4.onion – Tor Search
http://cipollatnumrrahd.onion/ – Cipolla 2.0 (Italian)

Fig. 4 TOR hidden services with their .onion URL .onion is a top-level domain designed for anonymous onion services

Moreover, a very small number of services are legitimate in TOR. Other services (drugs, human trafficking, etc.) are obfuscated to a large extent to rumble the TOR system and its users (Fig. 4) shows the TOR hidden services with their .onion URL.

Even if the TOR network is focused on privacy, the authorities managed to identify and hold responsible for illegal activities users of the network and initiators of hidden services. For example, Ross Ulbricht [1] is a very popular case, responsible for operating an online marketplace of drugs named "Silk Road" in 2013. As a result, he was arrested by the Federal Bureau of Investigation (FBI). Another similar case in 2015 is "Playpen", a distribution service for child pornography using the TOR network [2]. Jason Sebastian Sparks was responsible for operating "Playpen" was eventually sentenced to 14 years in prison, later followed by 15 years of supervised release. In 2019, a similar hidden service, "The Giftbox Exchange" [3] came to light where four men were sentenced to 30 years of prison for distributing illegal child pornography over an open-source TOR network.

The majority of illicit websites on the TOR network can be fully accessed using Bitcoin payments, which are anonymous transactions. However, there are other cryptocurrencies available, not popular as Bitcoin is. Recently in the USA, the police have seized down $613,600 in cryptocurrencies, which were used in the largest Dark Web market, "The Wall Street Market". Another famous case is Silk Road, where the FBI seized 26,000 Bitcoins worth around $3.6 million at that time. Furthermore, ransomware attacks break the backbone of various countries and raise a serious question on their security platforms, as the ransomware payment sites ask for Bitcoin as ransom, and it is pretty hard to catch the real attacker, as described in reference [4].

TOR gives you an opportunity to hide your identity, but it can reveal the content flowing through the network. Therefore, it is highly recommended to use nested security or layered security while hosting your hidden service. One way to do this is to use an anonymous server to host your website, but certainly, there are other flaws in the TOR network itself that can backtrack and can disclose the intention behind the hidden service. In reference [5], the authors discuss how TOR and services like Bitcoin can be used to anonymize user activities like web surfing and online transactions. They analyzed that Bitcoin lacks retroactive operational security, where we can fetch the historical pieces of information of users and hence can deanonymize the user. The indexed 1.5 K hidden services and gathered 88 unique and active Bitcoin addresses by analyzing the transactions. TOR is a marketplace of deadliest services such as narcotics and the distribution of illicit materials. Reference [6] shows the hidden service lifecycle and observes that previous estimates on the hidden service size are inaccurate and that the new estimates are found using lifecycle. Furthermore, the work analyzes the use of crawling and whether this is an effective mechanism to discover sites for law enforcement. Reference [7] focuses on existing attack schemes, their comparisons, the key concepts behind the attacks, and their interrelations. The paper shows a tabulated format to show each attack on TOR precisely and can be used as guidelines for further attack analysis. In addition to [7], reference [8] studies landscapes of TOR hidden services. The authors analyzed 39,824 hidden services gathered on February 4, 2013, scanned for open ports, scrutinized and classified their content, and estimated the popularity of hidden services by looking at the request rate of the service by the clients.

4 Installation of Hidden Service in Linux

To start with hidden service installation, we need the *Nginx* server. Though it is not mandatory to use this particular server, many alternatives such as Apache, Savant, and Wamp are available. Moreover, it is highly recommended not to use any server that gives potential information such as—port numbers and IP addresses as it can lead to disclosing your identity.

We start by installing the server using the following command.

$ sudo apt install nginx

The command above installed the nginx server in /etc./nginx folder.

We need to configure the firewall (ufw in this example) to allow HTTP traffic from Nginx.

$ sudo ufw allow 'Nginx HTTP'

$ sudo ufw reload

Then, we should configure the IP address and listening port numbers of the *Nginx* server. Finally, edit the default configuration file of Nginx, using the below commands.

```
$ sudo nano /etc/nginx/sites-available/default
```

Replace or add these lines of code to access the localhost and deny all other traffic.

```
server {
        listen 127.0.0.1:8080 default_server;
        server_name localhost;
        root /usr/share/nginx/html;
        index index.html index.htm;
        location / {
              allow 127.0.0.1;
              deny all;
        }
}
```

We completed our initials setup, and our server is up and ready, start the Nginx service using the following command:

```
$ sudo service nginx start
```

Then, open the default browser and type the http://127.0.0.:8080 or http://localhost:8080 address in the browser address box and, if all the things were correct, you would see the default index .html page as shown in Fig. 5.

The further steps will guide you through the TOR installation process according to the Linux distribution you are using.

In order to see your Linux configuration, use the following command and it will show you the release and codename, as can be observed in Fig. 6.

```
$ lsb_release –a
```

Fig. 5 nginx web server—used for installing hidden service on TOR. Image shows the welcome message of nginx which confirms the successful installation of web server

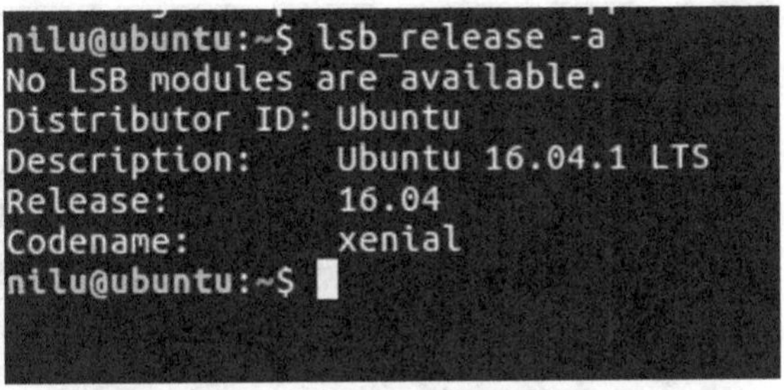

Fig. 6 Lsb_release – Linux standard Base reveals information about specific Linux distribution

```
nilu@ubuntu:~$ lsb_release -a
No LSB modules are available.
Distributor ID: Ubuntu
Description:     Ubuntu 16.04.1 LTS
Release:         16.04
Codename:        xenial
nilu@ubuntu:~$
```

We are working with Ubuntu 16.04 LTS with the xenial version. Based on that, we have set up a package repository to fetch to TOR. Visit this website—https://2019.www.torproject.org/docs/debian.html.en#ubuntuto download TOR and not TOR browser (don't confuse with tor browser), then add the following entries in the /etc./apt/sources .list. using the following command,

$ sudo gedit /etc/apt/sources.list

Add these entries to the sources .list:

deb https://deb.torproject.org/torproject.org stretch main
deb-src https://deb.torproject.org/torproject.org stretch main

Then continue with the following commands:

$ curl https://deb.torproject.org/torproject.org/A3C4F0F979CAA22CDBA
8F512EE8CBC9E886DDD89.asc | gpg –import
$ sudo su
# gpg --export A3C4F0F979CAA22CDBA8F512EE8CBC9E886DDD89 | apt-key add –
# exit
$ apt update
$ apt install tor deb.torproject.org-keyring

While exporting *gpg* command, you will get a root error, as the *gpg* cannot be executed without a root user. Use the *sudo su* command to become root and the *exit* command to come back to the non-root user, as presented in Fig. 7.

Later commands will update the release and install the TOR network, and you can use the start and the stop command to start and stop the TOR services.

Once TOR is installed, navigate in the Terminal to configure the *torrc* file and enable the hidden service directory with the following command,

$ sudo nano /etc./tor/torrc.

You will see the file contents, like the one provided below.

#HiddenServiceDir /var/lib/tor/hidden_service/
#HiddenServicePort 80 127.0.0.1:80

Replace the port number from 80 to 8080,

```
nilu@ubuntu:~$ curl https://deb.torproject.org/torproject.org/A3C4F0F979CAA22CDBA8F512EE8CBC9E8
  % Total    % Received % Xferd  Average Speed   Time    Time     Time  Current
                                 Dload  Upload   Total   Spent    Left  Speed
100 19665  100 19665    0     0   2900      0  0:00:06  0:00:06 --:--:--  3735
gpg: key 886DDD89: "deb.torproject.org archive signing key" not changed
gpg: Total number processed: 1
gpg:              unchanged: 1
nilu@ubuntu:~$ gpg --export A3C4F0F979CAA22CDBA8F512EE8CBC9E886DDD89 | apt-key add -
ERROR: This command can only be used by root.
nilu@ubuntu:~$ sudo gpg --export A3C4F0F979CAA22CDBA8F512EE8CBC9E886DDD89 | apt-key add -
ERROR: This command can only be used by root.
[sudo] password for nilu:
gpg: [stdout]: write error: Broken pipe
gpg: build_packet(2) failed: file write error
gpg: WARNING: nothing exported
gpg: key export failed: file write error
nilu@ubuntu:~$ sudo su
root@ubuntu:/home/nilu# gpg --export A3C4F0F979CAA22CDBA8F512EE8CBC9E886DDD89 | apt-key add -
OK
```

Fig. 7 gpg command—it establishes the secure communication between two parties. In order to import gpg keys, need to change to root (sudo) user

HiddenServiceDir /var/lib/tor/hidden_service/

HiddenServicePort 80 127.0.0.1:8080

Next step, you can modify index.html according to the service you want to provide to the TOR users. The HTML files resides at **/usr/share/nginx/html/index.html**.

Moving on to the final steps, we will restart our server and TOR network, then fetch the .onion address from the hidden_service file.

$ sudo service nginx restart

$ sudo service tor restart

$ sudo gedit /var/lib/tor/hidden_service/hostname

The last command will show you the .onion website address of your service, which can be copied and opened in the TOR Browser. The result is presented in Fig. 8.

mzqeebtgyjrkkdqj23flmymx35ddesyfu5wnxj3duswqhxs42gmfgaqd.onion

ⓘ 🔒 mzqeebtgyjrkkdqj23flmymx35ddesyfu5wnxj3duswqhxs42gmfgaqd.onion

Welcome to Tor Hidden Service Tutorial

This is tor hidden service

Thank you for using my service

Fig. 8 Welcome message of newly installed TOR hidden service. It has .onion url and not generic top-level domain (.com,.in). The service can only be accessible with TOR network

The way this hidden service is installed is not the best approach, as this service can be easily detected and can disclose the real owner. There are several forensics attempts on specific users where the memory reveals .onion address with the IP addresses. Apart from that, there are already several vulnerabilities available in the TOR network that we can exploit in order to divulge the identity of either the user or the hidden services. Reference [9] implements a timing-channel fingerprinting attack on a hidden service on the TOR network hosted on Apache web server. The fingerprinting is an additive channel encoded with Reed–Solomon code for reliable recovery. In 60 min, the authors leave around 36-bit fingerprint and reliably recovered. However, the main challenge that hinders is the packet delays caused by the encryption of an anonymous network. Table 2 describes possible attacks categorized as client-side, server-side, network-side, and generic attacks. Elaborate attacks can be found in [10].

Table 2 Categorization of TOR attacks

Client attacks	• Plugin-based attack • Torben attack • P2P information leakage • Induced TOR guard selection • RAPTOR (Routing Attacks on Privacy in TOR) • Unpopular port exploitation • Low-resource routing attack • Bandwidth estimation attack • Passive linking attack • FortConsult security attack • Practical congestion attack • Connection start tracking attack • Stream correlation attack • Intersection attack • Statistical disclosure attack
Server attacks	• Cell counting and padding • TOR cells manipulation • Caronte attack • Off path man-in-the-middle attack
Network attacks	• Bridge discovery • Denial of service • Sniper • AS awareness attack
General attacks	• Traffic analysis attack • Timing attack • Shaping attack

5   Countermeasures to Secure Your Own Hidden Service

(1) Host and Listen to Localhost Only

No one should reach your onion web application through the normal Internet. The web server you are using for hosting hidden services should only listen to 127.0.0.1, and it doesn't leak any information that can disclose anonymity. We are not using hidden services on normal Internet with an actual IP address because Shodan and Censys can scan it, or even Google might be trying to index the server and eventually the service.

(2) Always Disable the Directory Listing

The server that hosts your hidden website has a default directory and an index .html file. This directory and that file (default index .html) can reveal certain information (list of icons, various file types, server details) that leads to disclosing the information about your service. Therefore, it should always be disabled using the **autoindex off;** parameter in the configuration file for the Nginx server.

(3) Fingerprinting Data

Always disable the server fingerprinting data such as header, version number, cookies, and IP addresses that can track your service. In Apache server, you must disable server-info and server-status information by configuring *httpd .conf* file. In order to do this, you should comment out these two lines from *httpd .conf* file.

```
#<Location/server-info>
#<Location/server-status>
```

In addition to that, you should also remove the server signatures such as server name, server version number and operating systems, as these signatures leak in default webpage of 404 or 500 errors. The security .conf file is responsible for handling such configuration. You can find this file in the following path /etc./apache2/conf-enabled/security .conf. You should add the following two lines in the mentioned file:

```
ServerSignature Off
ServerTokens Prod
```

(4) Only use TOR Traffic

The only way your service is going to be hidden is by routing all your traffic through TOR and not by using any surface Internet. There are numerous examples where your actual data leaks. For example, if you are using e-mail in TOR and providing your legitimate and real e-mail ID, then there are chances that it logs your real IP address in the mail server, which later can disclose the identity.

References

1. Greenberg et al., Silk road creator ross ulbricht loses his life sentence appeal, in *WIRED*, 2019. [Online]. Available: https://www.wired.com/2017/05/silk-road-creator-ross-ulbricht-loses-life-sentence-appeal/
2. Govinfo.gov, [Online] (2019). Available: https://www.govinfo.gov/app/details/USCOURTS-caed-2_16-cr-00095/USCOURTS-caed-2_16-cr-00095-5
3. Owner of the child abuse site The Giftbox exchange sentenced to 35 years in Prison. Dark-netStats, *DarknetStats*, 2019. [Online]. Available: https://www.darknetstats.com/owner-of-the-child-abuse-site-the-giftbox-exchange-sentenced-to-35-years-in-prison/
4. C. Cimpanu, Tor-to-Web Proxy Caught Replacing Bitcoin Addresses on Ransomware Payment Sites, *BleepingComputer*, 2021. [Online]. Available: https://www.bleepingcomputer.com/news/security/tor-to-web-proxy-caught-replacing-bitcoin-addresses-on-ransomware-payment-sites
5. H. Jawaheri, M. Sabah, Y. Boshmaf, A. Erbad, Deanonymizing Tor hidden service users through Bitcoin transactions analysis. Comput. Security **89** (2019)
6. G. Owenson, S. Cortes, A. Lewman, The darknet's smaller than we thought: The life cycle of Tor Hidden Services. Digit. Investig. **27**, 17–22 (2018)
7. S. Nepal, S. Dahal, S. Shin, Deanonymizing schemes of hidden services in tor network: A survey, in *International Conference on Information Networking (ICOIN)*, 12–14 Jan. 2015, Cambodia, pp. 468–473
8. A. Biryukov, I. Pustogarov, F. Thill, R. Weinmann, Content and popularity analysis of tor hidden services, in *IEEE 34th International Conference on Distributed Computing Systems Workshops*, 2014. 30 June–3 July 2014, Madrid, Spain, pp. 188–193
9. B. Shebaro, F. Perez-Gonzalez, J. Crandall, Leaving timing-channel fingerprints in hidden service log files. Digit. Investig. **7**, S104–S113 (2010)
10. E. Cambiaso, V. Ivan, L. Patti, Darknet security: a categorization of attacks to the To network, in *ITASEC 2019 - Italian Conference on Cyber Security*, 2019

Chapter 5
Introduction to Digital Forensics

1 Introduction to Forensics

Traditionally, we had a computer connected with a small number of peripheral devices to perform a programmed task. The system was quite efficient in those days, and however, the computer systems were not that emergent field to study and to do research and development because of many factors such as the difficulty in connecting any hardware to its software, the equipment cost was very high, the need for a large storage area, and not many people knew the internal development of computing. But gradually, after the industrial revolution, there is a constant need for automation rather than manpower. Automated tasks need some programming instructions given to the computers to perform the task repetitively. These programming instructions represent programs written using specific syntax and vocabulary in the so-called programming languages, which enhanced our computing power with many technologies. The expansion of the World Wide Web (WWW) had introduced a level-up technology, which proliferated many other platforms in terms of software, website, and applications development. The rise of e-commerce and e-governance adopted by several nations also increased the risk of cybercrime, as such platforms stores a large quantity of personal information. The intuition behind any cybercrime is to accumulate personal information from the Internet, which the intruder can leverage as he can use your data to do some malicious activity or sell your personal information to the third party or black market. Such incidents are rising day by day, and it is a challenging task to stop. To settle down the risk, nations have demanded several experts who have a strong grip on computer systems and their underlying law enforcement protocols. These experts are "cyberforensics", [1] who holds remarkable knowledge on computer, mobile, and networks to investigate any cybercrime done on such platforms.

Ransomware is the most dominant attack vector in the category of malware, which encrypts and restricts access to the file, threatening with permanent data destruction, even if the ransom is paid. The global economy gets fall due to its massive attack on several large companies, government organizations, hospitals, and telecom sectors,

N. Dutta et al., *Cyber Security: Issues and Current Trends*, Studies in Computational Intelligence 995, https://doi.org/10.1007/978-981-16-6597-4_5

cost of around $11.5 billion in 2019 gets disbursed. Denial-of-service (DoS) and distributed denial-of-service (DDoS) attacks are the second most powerful attack in cybercrime, according to the 2019 cybercrime statistics [2]. DoS floods the request packets to the targeted system that a user can't access the actual service; generally, IoT devices become the target as they are small, low-computing power devices and widely adopted for various services. Hacking such a miniature device instead of hacking a computer system protected by several layers of security is more accessible. DoS attacks affected larger companies in China, the USA, and Australia from 2017 to 2020, faced DoS attacks as 50.43% in China, 25% in the USA, and 4.50% in Australia.

Malware attacks are the third most significant attack vector in cybercrime. Roughly 1 out of every 50 e-mails are spammed and intended to attack with any malware family. It is the most common medium to communicate with any users, although social media is also an add-on with e-mail; however, 92% of this malware gets spread using e-mail. In addition to that, phishing attacks are also the most common form of attacks, where you can trick the human brain into clicking and visiting malicious content. As described below, phishing attackers use human weaknesses such as greed, lust, empathy, or curiosity [3].

Greed can be seen as an assurance of getting something valuable if you do this. An example of greed exploitation for phishing attacks is the message below.

Your Mobile No. is selected as a winner of $1750 on Coke Promo. Go to jangifts.net to claim. Enter Ref: AU556634393. Helpline: info@mobilecola.co.uk.

Lust is widely used by attackers, where they provide an online dating system or a similar platform that is wide open and free to access the pictures and videos you just want to see.

Empathy represents an urge to help someone who is in need to have that help. It mainly affects the people who do charity, environmentalists—think about the environment or an activist in any organization such as NGO (Fig. 1).

Curiosity affects the people who are curious to know what is behind the phishing link, and they want to know what happens next. Below is an example of such a message (Fig. 2).

Due to such reasons, there is a constant need for forensics, which can analyze and defend before and after cybercrime. However, the central role of cyberforensics exists after a cybercrime happens.

2 Cyberforensic Process

Cyberforensic is a practice that involves investigation and analysis to identify, collect, and preserve the evidence from an electronic gadget such as a computer or a hard disk, which can be suspect for further investigation. It is the only suitable evidence that can be present in a court of law. Indeed, it is difficult to hold evidence for a longer time, as evidence is purely volatile and often hidden in layers of assembly language. It is not an easy task to get the flag from such massive information. Investigators use

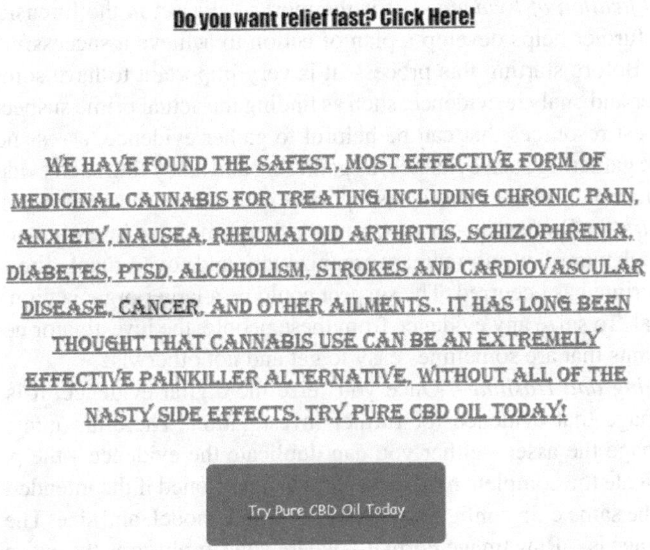

Do you want relief fast? Click Here!

WE HAVE FOUND THE SAFEST, MOST EFFECTIVE FORM OF
MEDICINAL CANNABIS FOR TREATING INCLUDING CHRONIC PAIN,
ANXIETY, NAUSEA, RHEUMATOID ARTHRITIS, SCHIZOPHRENIA,
DIABETES, PTSD, ALCOHOLISM, STROKES AND CARDIOVASCULAR
DISEASE, CANCER, AND OTHER AILMENTS.. IT HAS LONG BEEN
THOUGHT THAT CANNABIS USE CAN BE AN EXTREMELY
EFFECTIVE PAINKILLER ALTERNATIVE, WITHOUT ALL OF THE
NASTY SIDE EFFECTS. TRY PURE CBD OIL TODAY!

Try Pure CBD Oil Today

Fig. 1 Phishing e-mail comprises some help to cancer people. A user who has a desire to help will click on that "red" color button. It could be malware, clickbait, or something malicious

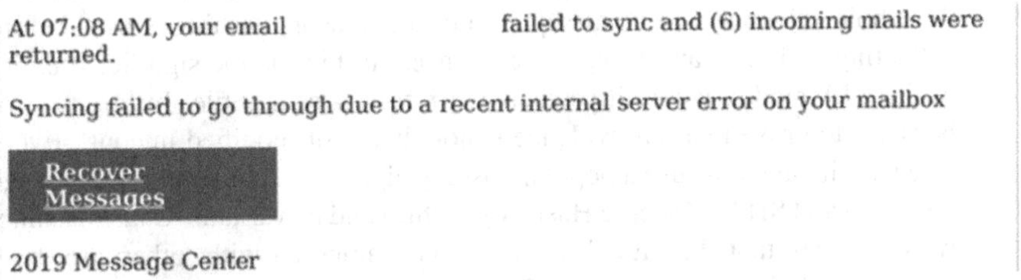

At 07:08 AM, your email   failed to sync and (6) incoming mails were returned.

Syncing failed to go through due to a recent internal server error on your mailbox

Recover
Messages

2019 Message Center

Fig. 2 A curiosity phishing mail consists of a recovery message. If users are curious to recover their previous messages, they will click on that "blue" button. [*Source* https://retruster.com/blog/phishing-email-scams-with-real-phishing-examples.html]

several techniques and applications to examine the collected evidence, which can be an image, encrypted files and folders, unallocated disk spaces, hidden folders, or any damaged files. The investigation process includes the following steps: Incident Spot, Identification of evidence, Seizure of evidence, Imaging and Hashing of evidence, Analysis of evidence, Reporting and Preservation of evidence till it could be presented in the court of law.

(a) *Incident spot*—It is the occurrence point where a cybercrime event took place and the devices like computers, mobile devices, hard drives, etc., that have been used to commit the crime.

(b) *Identification of Evidence*—It is the most vital asset in the forensics process, as it further helps develop a plan of action to achieve a successful investigation. Before starting this process, it is very important to have some scope to gather and analyze evidence, such as finding the actual prime suspect, what are the best resources that can be helpful to gather evidence, assessment of cost while gathering such evidence, as gathering and analyzing such evidence needs commercial software and a susceptible environment to preserve the asset.

(c) *Seizure of Evidence*—To prove any cybercrime, the investigator has to go through many law enforcement organizations and government officials to prove that crime has occurred. The suspect could be a larger organization or an individual. To seize any evidence from these people, the investigator needs search warrants that are sometimes easy to get and not otherwise.

(d) *Imaging and Hashing*—Once you seize the digital evidence, it is important to image that evidence for further investigation. There are numerous ways to image the asset—either you can duplicate the evidence—the process will duplicate the complete hard drive, or it can be cloned if the intended hard drive has the same chip configuration such as brand, model, and size. The other way to image is—Raw Image Format—where you simply copy the entire hard drive bit by bit. The extension of imaging file format could be .*dd* (disk dump) or .*e01* (encase evidence file) or .*aff* (advance forensic format). The image file contains all folders and files (deleted and not deleted) and image metadata. Care should be taken while creating the image as if a single bit is changed or modified, and the whole evidence will be disrupted. For that reason, hashing is used before collecting and after analyzing the evidence. Hashing is the significant entity of the CIA model, where it protects the *integrity* of any file. In forensics, it is useful to prove that our evidence is not altered or modified in court. Every forensics image must use a popular hashing algorithm such as MD5 (Message Digest 5) and SHA 1 (Secure Hash Algorithm) and its variants. Using hashing, we can know that the digital asset was not tampered with, otherwise, both images get different hash values (Table 1).

(e) *Analysis and Reporting*—This phase comes after the imaging and hashing. It further goes to the investigation, where the investigator looks for findings and outliers that can prove a crime happened and who is behind it. Reporting phase comes after a successful analysis of the cybercrime. All conclusions of the analysis phase should be presented in a report format by the investigator. The report should be in the formal structure, and there should not be any personal

Table 1 Different hashing algorithms, with their length of the digest

Hashing algorithm	Length of digest (Bit)
MD5	128
SHA-1	160
SHA-256	256
SHA-512	512

Fig. 3 Phases of digital cyberforensics process—Any cybercrime that occurs, a forensic agent will have to go through these phases to track and hunt the criminal behind the crime

views in the report. However, it should contain an in-depth analysis with a tangible conclusion that any non-technical person could understand (Fig. 3).

3 Different Artifacts and Forensic Tools

Investigation of any cybercrime has to go through a series of acquisition and analysis processes. There are simple steps to do the acquisition process: "RAM Capturing", turn off the system, seizing of HDD/SSD, perform imaging and hashing of the storage device, investigate and analyze the evidence and try to find out the suspect using various open-source and commercial tools (Fig. 4).

As an attacker uses hybrid tools and applications to do the malicious activity, it is really hard to find a needle (attacker) in the haystack of different technologies that they had used. Still, an agent has many ways to analyze, leading to the attacker, as every suspect leaves a piece of evidence in an incident spot. Below are some of the digital pieces of evidence which can be useful in the investigation process of any crime scene:

(1) Operating system artifacts that include files and folders (hidden)
(2) Hard drive (HDD/SSD) of the victim
(3) Computer memory (RAM, cache)
(4) Network traffic
(5) Any cloud storage—(Google Drive or One Drive, Dropbox)
(6) Mobile phone and tablets
(7) Network devices
(8) Log file—system logs, network logs, device logs
(9) GPS information (using Google Activity Control)
(10) Event Viewer.

To do an investigation, a team called IR (Incident Response) has to be formed. Its members possess certain skills and technical expertise that greatly helps in finding out

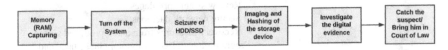

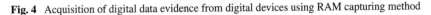

Fig. 4 Acquisition of digital data evidence from digital devices using RAM capturing method

the conclusion from the investigation. IR team should have personal and technical skills. When an expert is interacting with a non-technical person, for example, a judicial person, it is crucial how effectively and simply an IR team can make them understand the crime scenario. If they can't do that, there might be a misunderstanding or misinterpretation of the result, leading to a wrong conclusion. Authors of [4] have shown live forensics on software—they have used RAM artifacts of Java programs, they extracted program's state even when the garbage collector is called explicitly, the software is stopped, or JVM is closed. Authors of [5] have integrated machine learning algorithms into digital forensics to detect online sexual predatory chats. Researchers have used chat logs as digital footprint from social media to detect detrimental conversation using greedy search algorithms (Table 2).

The investigator must have a complex hardware and software kit for the investigation process. It depends on the organization's budget to have a set of equipment to draw out the conclusion from the digital evidence.

Table 2 Personal and technical skills required by an Incident Response team (IRT), which helps the non-technical person to understand the outline of crime

Personal skills	Technical skills

• Written and oral communication
• Effective presentation skills
• Diplomacy—maintain good relationships, and avoid offenders
• Follow policies and procedure
• Team skills
• Maintaining integrity between people
• Should know their limits
• Problem solving and a good observer
• Time management
	• Includes security principles-
	  – Confidentiality
	  – Availability
	  – Authentication
	  – Integrity
	  – Access control
	  – Privacy
	• Security weakness and vulnerability
	  – Physical security issue
	  – Protocol design flow
	  – Malicious code
	  – Implementation flaw (buffer overflow, race conditions)
	  – Weakness in configuration
	  – Human errors
	• The Internet and computer security risk
	• Understanding of law and criminal investigation
	• Willingness to adapt to new platforms and technologies
	• Network application and services
	• Host or system security issues
	  – Harden your system
	  – Review configuration file
	  – Measure common attacks and methods
	  – Review log files
	  – Review system privileges
	  – Secure network daemons
	  – Recovery measures from a compromised system
	• Programming skills—C, Python, Shell, Java, and other scripting tools

The following hardware items should be present in any investigator kit:

(1) Jump Bag—FREDL + UltraKit (Forensic Recovery of Evidence Device Laptop)
(2) Network cables
(3) Several serial cables and USB adapter
(4) Network serial adapter
(5) Hard drives (HDD/SSD)
(6) Flash drive
(7) Digital camera
(8) Notebook, pen
(9) Cable snips
(10) Linux Live DVD
(11) Portable drive duplicator
(12) Write blocker
(13) Screw and hex drivers and port hub
(14) Custody forms.

The software kit deals with collecting volatile and non-volatile data. Numerous software solutions in the market with both command-line utilities and GUI applications can be useful to the particular analysis.

Volatile Data represents any data stored inside the computer memory and will be lost when the computer is turned off or loses power. Such information is beneficial in forensics to draw out the conclusion. Below are some of the important entities which can be gathered from memory analysis.

(1) System uptime and current time
(2) Network connections parameters (NetBIOS name, cache, active connections, routing table, and so on)
(3) Network interface card (NIC) configuration settings
(4) Logged on users and active sessions
(5) Drivers
(6) Running process and services
(7) Autostart handlers
(8) Shared and remote files and drives (Table 3).

Table 3 Open-source and commercial tool to dump the RAM as a digital asset for investigation

Open-source tools	Commercial tools
DumpIt	F-response
DEFT (open-source operating system)	Guidance Winen
WinPMEM	HBGary Fastdump PRO
Belkasoft Live RAM capturer	X-Way Forensics
Xplico for network analysis	Forensics toolkit
Sleuth kit	Encase

You can gather much information from a running system by using sysinternals tools. Sysinternal suite can be downloaded from https://docs.microsoft.com/en-us/sysinternals/downloads/sysinternals-suite and contains utilities such as nstat, promqry, psloggedon, and so on. Reference [6] explains how sysinternals tools could be helpful in various administrative tasks for any computer. Furthermore, it considers how to customize, configure and automate the task using command-line parameters.

The following tools and commands from the sysinternals suite are useful for any investigation:

1.   *nbstat-c*—obtain information from the cache of NetBIOS
2.   *netstat –arn*—record all of the active compounds, listening ports, and routing tables
3.   *ipconfig/all*—network interface command
4.   *Promqry*—network interfaces on the local machine, which operates on promiscuous mode
5.   *psloggedon*—lists both types of users, logged on to the computer locally and logged on remotely over the network. ($-x$) parameter for timing
6.   *logonsessions*—it will display all the processes that are executed by the user and are active
7.   *tasklist*—it will enumerate the list of running processes and services
8.   *listdlls*—it enumerates a list of dll (dynamic link library) in a system
9.   *handle.exe*—it will list all handles such as—registry key, ports, mutexes, etc.
10. *autorunsc.exe*—list of .exe that run at system startup or bootup.

In the following sections, we will present the fundamental characteristics of the tools needed for digital forensic.

3.1 Autopsy

Is a digital forensic tool to generate memory images of the Windows operating system. It comes up with a nicely suited graphical interface by which we can create a memory dump of the entire hard drive or any specific partition of the drive. Many of its programs can be used as an add-on in The Sleuth Kit, which is an underlying tool used by many forensic investigators to analyze certain sections of data in a detailed manner. It can analyze various file systems, such as NTFS, FAT, ExFAT, Ext2/Ext3/Ext4, and many more. Autopsy can be downloaded from https://www.sleuthkit.org/autopsy/. It can be installed using the default setting, and once opened, the user can choose the data source memory file that exists on their disk or create a memory dump of the hard disk for later analysis.

Reference [7] had given a systematic comparison and analysis between commercial and open-source mobile device forensic tools, where commercial tools—MOBILEedit! Forensics, Cellebrite's UFED physical analyzer, and open-source

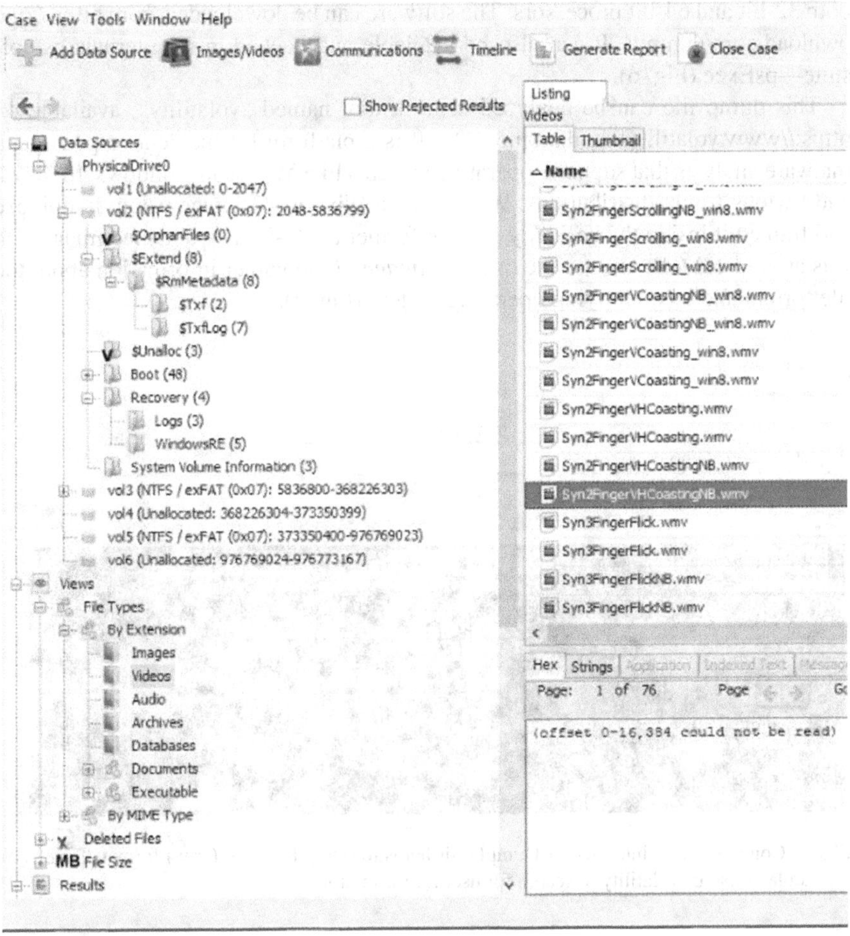

Case View Tools Window Help

Add Data Source   Images/Videos   Communications   Timeline   Generate Report   Close Case

☐ Show Rejected Results

Listing
Videos

- Data Sources
  - PhysicalDrive0
    - vol1 (Unallocated: 0-2047)
    - vol2 (NTFS / exFAT (0x07): 2048-5836799)
      - $OrphanFiles (0)
      - $Extend (8)
        - $RmMetadata (8)
          - $Txf (2)
          - $TxfLog (7)
      - $Unalloc (3)
      - Boot (48)
      - Recovery (4)
        - Logs (3)
        - WindowsRE (5)
      - System Volume Information (3)
    - vol3 (NTFS / exFAT (0x07): 5836800-368226303)
    - vol4 (Unallocated: 368226304-373350399)
    - vol5 (NTFS / exFAT (0x07): 373350400-976769023)
    - vol6 (Unallocated: 976769024-976773167)
  - Views
    - File Types
      - By Extension
        - Images
        - Videos
        - Audio
        - Archives
        - Databases
        - Documents
        - Executable
      - By MIME Type
    - Deleted Files
    - MB File Size
  - Results

Table | Thumbnail

▵ Name

- Syn2FingerScrollingNB_win8.wmv
- Syn2FingerScrolling_win8.wmv
- Syn2FingerScrolling_win8.wmv
- Syn2FingerVCoastingNB_win8.wmv
- Syn2FingerVCoastingNB_win8.wmv
- Syn2FingerVCoasting_win8.wmv
- Syn2FingerVCoasting_win8.wmv
- Syn2FingerVHCoasting.wmv
- Syn2FingerVHCoasting.wmv
- Syn2FingerVHCoastingNB.wmv
- Syn2FingerVHCoastingNB.wmv
- Syn3FingerFlick.wmv
- Syn3FingerFlick.wmv
- Syn3FingerFlickNB.wmv
- Syn3FingerFlickNB.wmv

Hex | Strings | Application | Indexed Text | Message

Page: 1 of 76   Page ← →   Go

(offset 0-16,384 could not be read)

Fig. 5 Interface of Autopsy digital forensic platform—Data Source has all drives, which will be used for investigation. Narrow down your result using file types. The relevant content of files can be extracted from Hex or strings are shown in the bottom right

tools such as—The Sleuth Kit (with autopsy) and SANS SIFT (SANS Investigative Forensic Toolkit)—serve all the purposes of forensic investigation in Ubuntu platform (Fig. 5).

3.2 DumpIt

Is a utility for Microsoft Windows to create a physical memory dump of any Windows operating system. It is a freeware software developed by Moonsols and supports both 32-bit and 64-bit processors. The software can be downloaded from https://qpd ownload.com/dumpit/. It can also be available and packed in sysinternals pstools suite—psExec (Fig. 6).

This dump file can be analyzed using a tool named "volatility", available on https://www.volatilityfoundation.org/26. It is a platform for incident response and malware analysis that supports operating systems like Microsoft Windows, Mac OS, and various Linux distributions. We created a Windows 7 image using dumpit.exe and framed it inside the *volatility* tool for further analysis. The obtained image size was around 1.2 GB. We use the option—*imageinfo* to gather information about the file format and its underlying operating system (Fig. 7).

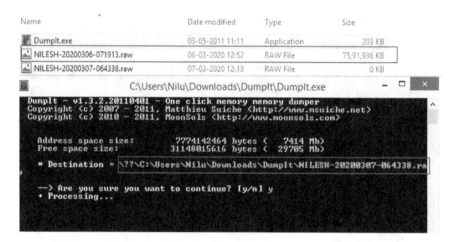

Fig. 6 Command-line interface of DumpIt—it helps to dump RAM in (.raw) format. This RAW file can later use in Volatility to search for useful information

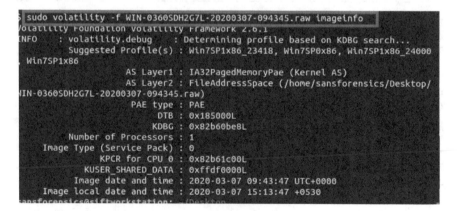

Fig. 7 Volatility tool—shows the profile information using—"imageinfo" options. It uses KDBG (kernel debugger), which will tell version information of the operating system

Volatility will process the image file (.raw) according to the image size. As this image is 1 GB, it took around 10 min to analyze it and show the results. However, it will take a lot of time to process it if the size of the dump is large. It also depends on the system configuration on which you run the tool, like processing power and RAM. The tool has provided us with information about the user using Windows 7 profile, specifically 32-bit operating system, number of processors allotted—1, and timing when the image is created.

$ sudo volatility -f WIN 0360SDH2G7L-20200307-094345.raw --profile=Win7SP1x86_23418 pslist

This will show you all the processes running in the Windows 7 system when the crime occurred, as shown in Fig. 8.

```
$ sudo volatility -f WIN-0360SDH2G7L-20200307-094345.raw --profile=Win7SP1x86_2
3418 pslist
Volatility Foundation Volatility Framework 2.6.1
Offset(V)   Name                    PID   PPID   Thds    Hnds   Sess  Wow64 Sta
rt                          Exit
---------- --------------------- ------ ------ ------ -------- ------ ------ ---

0x8413aa20 System                     4      0     83      522 ------        0 202
0-03-07 09:42:10 UTC+0000
0x85629b80 smss.exe                 268      4      2       29 ------        0 202
0-03-07 09:42:10 UTC+0000
0x85d9bd40 csrss.exe                344    336      9      769      0        0 202
0-03-07 09:42:14 UTC+0000
0x85da7438 wininit.exe              396    336      3       74      0        0 202
0-03-07 09:42:14 UTC+0000
0x85fbad40 csrss.exe                408    388     11      218      1        0 202
0-03-07 09:42:14 UTC+0000
0x85fbfd40 winlogon.exe             456    388      4      107      1        0 202
0-03-07 09:42:15 UTC+0000
0x86331030 services.exe             500    396      8      239      0        0 202
0-03-07 09:42:15 UTC+0000
0-03-07 09:42:18 UTC+0000
0x864cd4f0 spoolsv.exe             1280    500     13      332      0        0 202
0-03-07 09:42:19 UTC+0000
0x864e9220 svchost.exe             1316    500     22      324      0        0 202
0-03-07 09:42:19 UTC+0000
0x8652f970 httpd.exe               1420    500      7      142      0        0 202
0-03-07 09:42:20 UTC+0000
0x8653bc48 FileZilla Serv          1468    500      8       59      0        0 202
0-03-07 09:42:20 UTC+0000
0x8655b030 RaRegistry.exe          1532    500      5       81      0        0 202
0-03-07 09:42:21 UTC+0000
0x8656b030 xampp_service_          1556    500      6       97      0        0 202
0-03-07 09:42:21 UTC+0000
0x86566d40 mysqld.exe              1580    500     21    35646      0        0 202
0-03-07 09:42:21 UTC+0000
0x86566768 mercury.exe             1588   1556     17      130      0        0 202
```

Fig. 8 Volatility—pslist—It exhibits all the processes which were running on OS. The timestamp of the processes will tell you the important processes which are used in the crime

```
firefox.exe pid:    1564
Command line : "C:\Users\Nilu\Desktop\Tor Browser\Browser\firefox.exe"


Base            Size   LoadCount Path
----------   ----------  ----------  ----
0x01070000    0x16c000    0xffff  C:\Users\Nilu\Desktop\Tor Browser\Browser\firefox.exe
0x77ae0000    0x13c000    0xffff  C:\Windows\SYSTEM32\ntdll.dll
0x77860000    0xd4000     0xffff  C:\Windows\system32\kernel32.dll
0x75cb0000    0x4a000     0xffff  C:\Windows\system32\KERNELBASE.dll
0x64ae0000    0xe7000     0xffff  C:\Users\Nilu\Desktop\Tor Browser\Browser\mozglue.dll
0x716c0000    0xeb000     0xffff  C:\Windows\system32\dbghelp.dll
```

Fig. 9 Suspect using TOR browser can be extracted using volatility tool option—dlllist

From the analyses, we can observe that the FTP server is started along with MySQL and xampp. We can conclude that the crime is related to uploading or downloading something with FTP. The system has many vulnerabilities—mercury.exe (used to capture keyboard and mouse inputs), which can be exploited using Metasploit or any similar framework. We can analyze tons of information from this memory dump with our volatility tool as if you use *dlllist*, and the tool will gather information of all executables and show the specific *dll* that is appended with each executable. We can observe from Fig. 9 that the suspect is using TOR browser to hide its identity with process ID—1564.

$ *volatility_2.6_win64_standalone.exe -f WIN-0360SDH2G7L-20200312-051741.raw --profile=Win7SP1x86_23418 dlllist*

3.3 Belkasoft Live RAM Capturer

Belkasoft Live RAM capturer is an open-source tool for volatile memory acquisition. It extracts the entire content of the computer's volatile memory in *.mem* file format. The live memory can be analyzed using Belkasoft Evidence Centre. However, it is not free, though you can purchase it from here—(https://belkasoft.com/ec). You can create more dump files using process explorer, as it has options to have minidump and full dump (.dmp) of any specific process. The .dmp file can be analyzed using any debugging application such as WinDbg, Visual Studio, or DumpChk. The authors of [8] discuss several forensic tools for memory acquisition and the important features helpful in understanding any specific tool's pros and cons.

Flow and usage of online information are increasing, triggering new malware families who are unknown and hard to detect. Machine learning and computer vision have a good impact on cybersecurity, and incorporating them in forensics can be useful to detect such anonymous malware. In reference [9], the authors have captured the memory dump of illicit processes and converted them into RGB images

Name	Date modified	Type	Size
20200311.mem	11-03-2020 08:11	MEM File	75,91,936 KB
msvcp110.dll	22-10-2018 10:11	Application extens...	646 KB
msvcr110.dll	22-10-2018 10:11	Application extens...	830 KB
RamCapture64.exe	22-10-2018 10:11	Application	58 KB
RamCaptureDriver64.sys	22-10-2018 10:11	System file	34 KB

Belkasoft Live RAM Capturer

Select output folder path:

C:\Users\Nilu\Downloads\RamCapturer\x64

Loading device driver ...
Physical Memory Page Size = 4096
Total Physical Memory Size = 7414 MB
Memory dump canceled. Total memory dumped = 546 MB

Capture!   Cancel   Close

Fig. 10 Belkasoft Live RAM capturer will dump the whole drive memory in (.mem) format (20200311.mem ~ 7 GB). This dump can be given to Autopsy or Belkasoft evidence center for further investigation

using different rendering schemes. The dataset has 4294 samples containing 10 malware families. They got 96.39% accuracy by applying j48, RBF kernel bases SMO, Random Forest, XGBoost, and Linear SVM (Fig. 10).

4 Artifacts Gathering

There are many places where a cyberattacker left his footprint, and the investigators can analyze different locations to find these footprints to capture the intruder. Below are such locations where we can find many such footprints to gather information about the real accused.

4.1 Browser Artifacts

Are the vital part of any forensic investigation. We can get artifacts such as search history, saved passwords, extensions, cookies and login, though they are in JSON format, difficult to analyze. Still, some tools can show tons of information from such artifacts (Fig. 11, Tables 4, 5).

Table		SELECT * FROM moz_inputhistory	
⊞ moz_anno_attributes ‹			
⊞ moz_annos ‹			Input Keywords used
⊞ moz_bookmarks ‹			
⊞ moz_bookmarks_de.... ‹	I Place_id	Input	Use_count
⊞ moz_historyvisits ‹	29218	https://www.studyinindia.gov.in	0.12654605680588246
⊞ moz_hosts ‹	29149	scre	0.1495885731616798
⊞ moz_inputhistory ‹	29384	igno	0.18067631244528579
⊞ moz_items_annos ‹			

Fig. 11 Sqliteviewer—we can place a.json file to inspect what is the browser history, downloads, or profile information

Table 4 Firefox browser artifacts with their file path. Information such as browser history, ad-on, permissions, cookies, and session information can be extracted

Firefox browser artifacts	File path
Browser open and closed history for each session	..\bo72m8r2.default\datareporting\session-state.json
Browser crash history	..\bo72m8r2.default\crashes\store.json.mozlz4
Browser extension and ad-on	..\bo72m8r2.default\extensions\(…).xpi
Browser login and password	..\bo72m8r2.default \login.json
Browser session data	..\bo72m8r2.default \sessionstore.jsonlz4
Favicons	..\bo72m8r2.default\favicons.sqlite
Browsing history	..\bo72m8r2.default\places.sqlite
Granted permission for various sites to access camera, microphone, notifications	..\bo72m8r2.default\permissions.sqlite
Browser cookies	..\bo72m8r2.default\cookies.sqlite
HSTS information (SuperCookie) information	..\bo72m8r2.default\SiteSecurityServiceState.txt

Table 5 Chrome browser artifacts with their file path. Information such as browser history, extensions, permissions, cookies, form history, last tab, and session information can be extracted

Chrome browser artifacts	File path
Browser extension and ad-on	…\default\Extensions
Browser login and password	…\default\Login Data
Browser session data	..\default\Session Storage
Favicons	…\default\Favicons
Browsing history	…\default\History
Last tab opened	..\ …\default\Last Tabs
Browser cookies	..\default\Cookies
Form history	..\default\Web Data

For Firefox –

> %UserProfile%\AppData\Roaming\Mozilla\FireFox\Profiles\<...>.default
> \places.sqlite—will store Firefox annotations, bookmarks, icons, search history, keywords, and browsing history (websites of regular usage).
> %UserProfile%\AppData\Roaming\Mozilla\FireFox\Profiles\<...>.default
> \login.json—with login.json and key3 and key4.db can be helpful to get the username and passwords of the users.
> %UserProfile%\AppData\Roaming\Mozilla\FireFox\Profiles\<...>.default
> \sessionstore.jsonlz4—will give you the sessions of the users.
> %UserProfile%\AppData\Roaming\Mozilla\FireFox\Profiles\<...>.default
> \downloads.json—will provide you with a downloaded files list from the Firefox browser.

4.2 Registry Artifacts

Is a salient place where the Windows system stores most of its configuration. It acts as a database that stores every minute detail of the system, and it is modified every second. A forensic team will acquire a copy of the registry or they can inspect the hive to find crucial results.

RunMRU—It is a registry that saves all the recent commands from the Run dialog box. This could be helpful if a suspect is fond of shortcuts, he will directly execute the command from the Run dialog box. We can grab many footprints such as ping command, accessed files, folders or network share folders. The Registry Access path is \HKEY_CURRENT_USER\Software\Microsoft\Windows\Current-Version\Explorer\RunMRU (Fig. 12).

OpenSavedPidMRU is a registry file where we can access the recent open and closed process from the Open and Save file dialog boxes. You can find various extensions (e.g., .txt, .html, .pdf, .css) that you had opened and closed recently. This

ab g	REG_SZ	%temp%\1
ab h	REG_SZ	ping 192.168.15.91 -t\1
ab i	REG_SZ	ping 192.168.137.1\1
ab j	REG_SZ	ping 192.168.43.201\1
ab k	REG_SZ	ipconfig\1
ab l	REG_SZ	notepad++\1
ab m	REG_SZ	gpedit.msc\1
ab MRUList	REG_SZ	dgyfpalchwvteusxkmbqoinrjz
ab n	REG_SZ	node\1

Fig. 12 RunMRU—It will draw out the information about the commands executed in the Run dialog box. It can be observed that the user ping some IP addresses and also opened notepad++

is a precise analysis where we can explore what kind of files and specifically what files the suspect had opened or accessed.

The registry path is HKEY_CURRENT_USER\Software\Microsoft\Windows\CurrentVersion\Explorer\ComDlg32\OpenSavedPidMRU. These are the files that are saved inside the open and save file dialog and the same thing can be accessed using a mini tool—OpenSaveFilesView by Nirsoft, which can be downloaded from https://www.nirsoft.net/utils/open_save_files_view.html (Fig. 13).

Bags and BagMRU stores the files and folders recently browsed by the user. In Fig. 14, we can see that BagMRU had grabbed the IP address—172.21.16.236 that someone had accessed for file-sharing purposes. The path for Registry

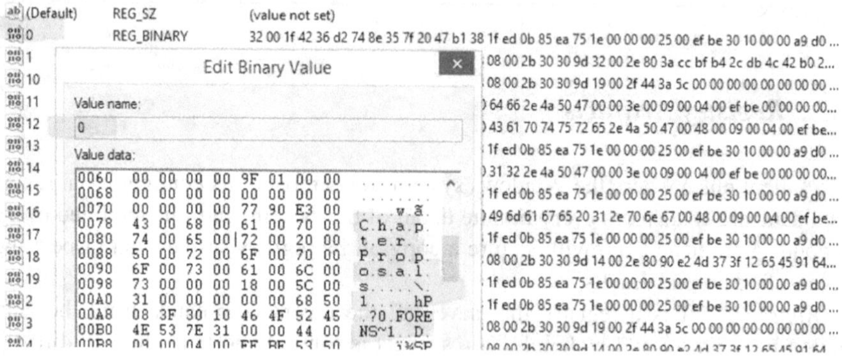

Fig. 13 OpenSavedPidMRU—registry file will explore the most recent files that have been saved or opened. This figure reveals that "chapter proposal" is the file the accused had recently opened

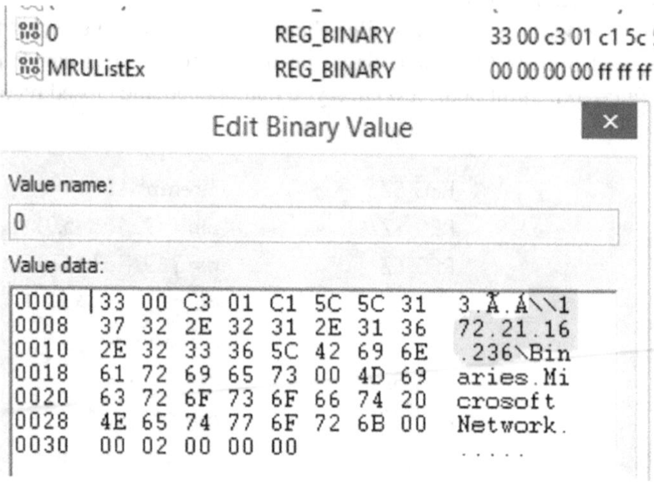

Fig. 14 Bags and BagMRU—registry file brings out the IP address of the file which is shared

Name	Type	Data
ab (Default)	REG_SZ	(value not set)
DefaultGatewayMac	REG_BINARY	20 39 56 6c 6e 4e
ab Description	REG_SZ	Tp-Link (Open wifi)
ab DnsSuffix	REG_SZ	<none>
ab FirstNetwork	REG_SZ	Tp-Link (Open wifi)
ab ProfileGuid	REG_SZ	{5C005E6C-5DF9-42E0-9F5D-1AAECCF57E07}
Source	REG_DWORD	0x00000008 (8)

ab (Default)	REG_SZ	(value not set)
{E2E0EBA8-817C-4A83-85B2-9F345D4BCBDD}	REG_BINARY	00 19 e8 f8 26 47
Failures	REG_DWORD	0x94040100 (2483290368)
Successes	REG_DWORD	0x6bfbfeff (1811676927)

Fig. 15 Network registry—registry will extract information such as managed or unmanaged network. Highlighted markings show us default MAC, SSID name

Access is Computer\HKEY_CURRENT_USER\Software\Microsoft\Windows\Shell\ BagMRU.

Network Artifacts is a registry that collects the information about the network where you connected—wired or wireless, identifies SSID, MAC addresses, domain name, and Intranet information. Registry Access— Computer\HKEY_LOCAL_MACHINE\SOFTWARE\Microsoft\Windows NT\CurrentVersion\NetworkList\Signatures\Unmanaged (Fig. 15).

USB and USBSTOR registry will gather information about the USB devices which are connected to the system, information such as—device name, driver information, and hardware ID. In addition, it identifies the USB vendor name, version number, the time when a device was plugged in, and the serial number of the USB or media device. The Registry Access path is HKEY_LOCAL_MACHINE\SYSTEM\CurrentControlSet\Enum\USBSTOR (Fig. 16).

Portable Devices is the registry where all the portable media attached to the computer, its drive name, and letter can be accessed. The Registry Access path for it is HKEY_LOCAL_MACHINE\SOFTWARE\Microsoft\Windows Portable Devices\Devices.

Plug and Play Driver (PnP)—When a PnP device is installed, the system will log with an event ID 20001. It includes devices such as—USB, PCMCIA, or any Network Interface Card, but not limited to. The log can be accessed on the path -/Windows/System32/winevt/logs/System.evtx, as shown in Fig. 17.

The registry entry "HKEY_USERS\S-1-5-21–520884347-4248758235-1010766 732-1001\Software\Microsoft\Internet Explorer\LowRegistry\Audio\PolicyConfig\ PropertyStore\606e4acd_0" will show the TOR browser installation path and its audio settings (Fig. 18).

HKEY_CLASSES_ROOT is a registry key containing all the extensions that a system uses, such as .mp3, .png, .pdf. This key defines the default program that is

ab ClassGUID	REG_SZ	{4d36e967-e325-11ce-bfc1-08002be10318}
ab CompatibleIDs	REG_MULTI_SZ	USBSTOR\Disk USBSTOR\RAW GenDisk
ab ConfigFlags	REG_DWORD	0x00000000 (0)
ab ContainerID	REG_SZ	{1a63a807-5215-5885-9172-41edbcceb697}
ab DeviceDesc	REG_SZ	@disk.inf,%disk_devdesc%;Disk drive
ab Driver	REG_SZ	{4d36e967-e325-11ce-bfc1-08002be10318}\0003
ab FriendlyName	REG_SZ	SanDisk Ultra USB Device
ab HardwareID	REG_MULTI_SZ	USBSTOR\DiskSanDisk_Ultra_____1.00 USBSTOR'
ab Mfg	REG_SZ	@disk.inf,%genmanufacturer%;(Standard disk drives)

ab ContainerID	REG_SZ	{02087ebc-58fc-5616-a5d0-6aefea4a0ae7}
ab DeviceDesc	REG_SZ	SM-A305F
ab Driver	REG_SZ	{eec5ad98-8080-425f-922a-dabf3de3f69a}\0001
ab FriendlyName	REG_SZ	Galaxy A30
ab HardwareID	REG_MULTI_SZ	USB\VID_04E8&PID_6860&REV_0400&MS_COMP_N
ab LowerFilters	REG_MULTI_SZ	WinUsb
ab Mfg	REG_SZ	Samsung Electronics Co., Ltd.

Fig. 16 USBSTOR—provides information such USB name (SanDisk), Phone name—Galaxy A30, Manufacturer name—Samsung and Device description—Version: SM-A305F (Global)

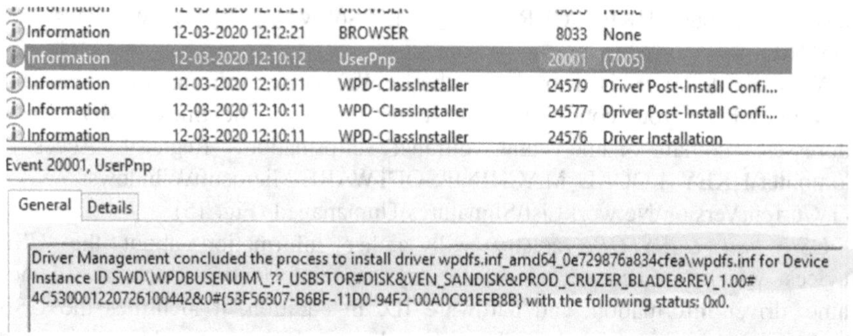

ⓘ Information	12-03-2020 12:12:21	BROWSER	8033	None
ⓘ Information	12-03-2020 12:12:21	BROWSER	8033	None
ⓘ Information	**12-03-2020 12:10:12**	**UserPnp**	**20001**	**(7005)**
ⓘ Information	12-03-2020 12:10:11	WPD-ClassInstaller	24579	Driver Post-Install Confi...
ⓘ Information	12-03-2020 12:10:11	WPD-ClassInstaller	24577	Driver Post-Install Confi...
ⓘ Information	12-03-2020 12:10:11	WPD-ClassInstaller	24576	Driver Installation

Event 20001, UserPnp

General   Details

Driver Management concluded the process to install driver wpdfs.inf_amd64_0e729876a834cfea\wpdfs.inf for Device Instance ID SWD\WPDBUSENUM_??_USBSTOR#DISK&VEN_SANDISK&PROD_CRUZER_BLADE&REV_1.00# 4C530001220726100442&0#{53F56307-B6BF-11D0-94F2-00A0C91EFB8B} with the following status: 0x0.

Fig. 17 Event viewer is an important utility in Windows system which logs information about system, application, security, and setup. It has generated an event with ID 20001 when Sandisk USB gets connected to the system

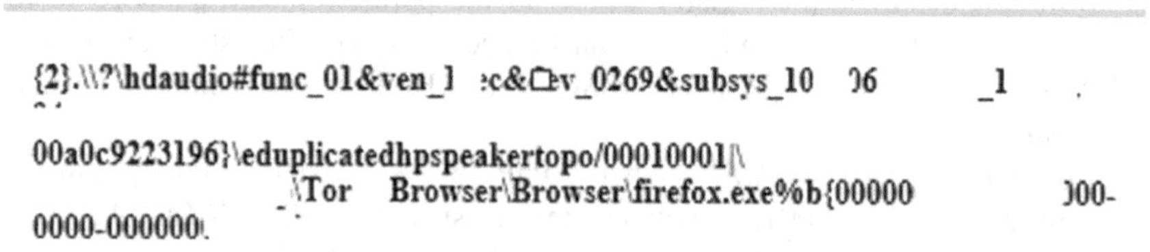

{2}.\\?\hdaudio#func_01&ven_] :c&C2v_0269&subsys_10)6   _1

00a0c9223196}\eduplicatedhpspeakertopo/00010001]\
          \Tor   Browser\Browser\firefox.exe%b{00000   )00-
0000-000000.

Fig. 18 HKEY_USERS saves active user-specific information. Here it can be seen that the suspect has installed the TOR browser, the username, the install location and some audio settings

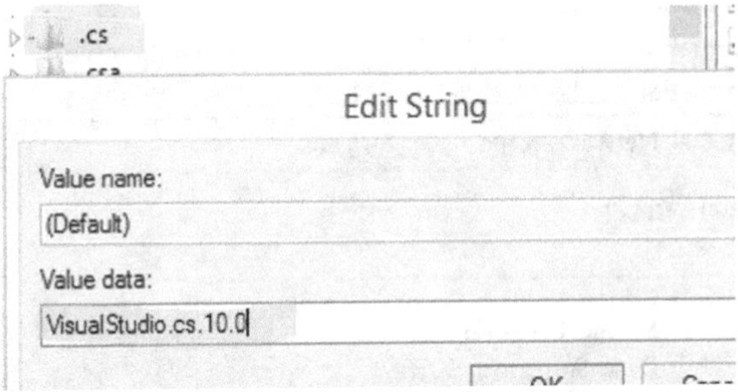

Fig. 19 HKEY_CLASSES_ROOT—registry file where all extension is stored, where an investigator can find what type of applications were used by the user

used to open files with a given extension. This information could be helpful for any investigator who can know the specific application used by the suspect for a given file type. For example, if we have a .cs extension, we can get to know that the user has Visual Studio 10 as an executable that supports the .cs extension (Fig. 19).

However, care should be taken, as Notepad++ or Sublime can also support .cs along with Visual Studio. The default program that the operating system will open for you is stored inside HKEY_LOCAL_MACHINE\SOFTWARE\Classes, but when you explicitly go for an alternative program to open the .cs file, it is stored inside HKEY_CURRENT_USER\SOFTWARE\Classes.

Event Log is an important utility from the Windows operating system, which stores application logs, new setup, system, and security. You can browse to Event Log by searching "View Event Logs". You can also navigate to \....\Windows\System32\winevt\Logs to find browser, network, or system-specific logs. For example, if the user connects to a wired or wireless network, the event can be logged inside Application and Service Logs—Network Profile, as shown in Fig. 20 (Table 6).

4.3 Bulk Extractor

Is a forensic tool that scans a digital memory image, directory of files of different formats (.raw, .mem), and extracts vital information without parsing any file structure. It has two versions—Bulk Extractor and Bulk Extractor Viewer—the first tool can take any digital image and will extract all the useful information and save in a single text file according to their domain (ip.txt, domain.txt, tcp.txt, aes.txt, and so on), in addition to that it also creates one .xml file called—Report.xml, which is a shorthand file to see all this information visually in a graphical format inside Bulk Extractor Viewer. For example, Fig. 21 shows the result of Windows 7 memory dump analysis, where we identify an onion URL from the TOR network.

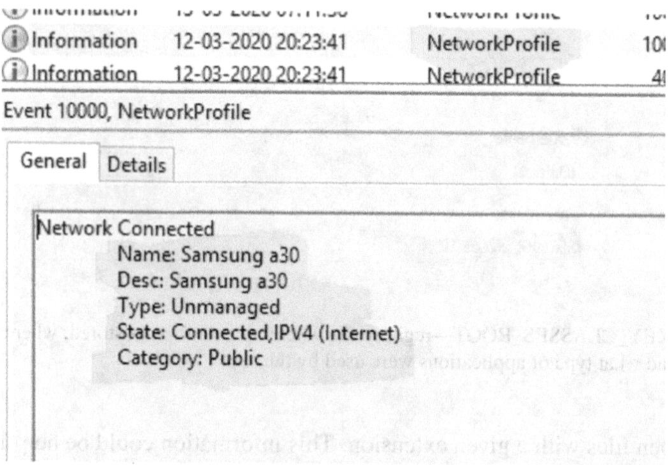

Fig. 20 Event viewer (Application and Service Logs)—indicates wireless hotspot as unmanaged network along with network name—Samsung A30 and its state—Public (connected)

Table 6 Windows event viewer log information

Logs	Information saved inside logs
System log	Account logon, logon events, policy group modifications, system events, account management, access control list privilege management
Security log	Logon activities, anonymous logon, uneven reboots
Directory service log	Logs of active directory
Application log	Logs of applications such as failure, modification to any application, malware activities with any application, antivirus disruption
DNS server	Logs of domains, zone transfer, management of zones, connected DNS server

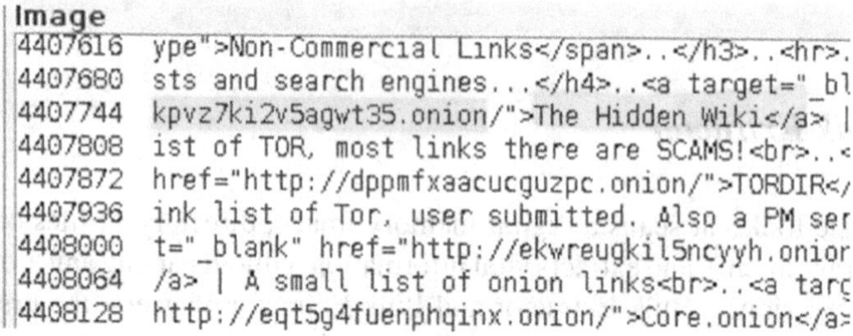

Fig. 21 TOR network hidden wiki (Onion URL)—extracted in bulk extractor viewer

Tables 7 and 8 present an overview of the registry tools, respectively, Windows forensic tools and their description.

Table 7 Registry tool to investigate registry in windows with its description

Registry tool	Description
RegSeeker	Small utility to search specific registry key
EncryptedRegView	A tool that scans any encrypted registry key and tries to decrypt it and display the data
RegShot	It quickly takes a snapshot of the registry before any specific instant of time and compares it with another snapshot
Registry ripper	It collects the hive files to extract information such as passwords from SAM files
USBOblivion	A tool that removes any trace of USB-connected storage devices from the registry
RegScanner	A utility that allows scanning all the registry from the system
RegFromApp	A tool that shows you the registry changes made by the application that you had selected
ActiveXHelper	Allow you to see vital information about any ActiveX Component
MiTeC Windows registry recovery	Registry analysis program

Table 8 Other forensics tools for windows operating system

Windows forensics tool	Description
WinPrefetchView	Prefetch files are the programs to speed up the program starting process
ThumbCache viewer	A tool that enables you to analyze thumbnail images
Exiftool	Allow you to see different information of any file such as—timestamps, size, geographic location of an image, if any
RBCmd	Provides you the artifacts from Recycle Bin
LastActivityView	Allows you to view what activities were taken by the users on that particular machine
HxD	Allows you to view the contents of a raw disk or main memory
Wireshark	It performs network forensic to see the network packets and the content inside visually
DCoDE	It calculates the date/time from the various timestamp of the data

(continued)

Table 8 (continued)

Windows forensics tool	Description
IEHistoryView	A tool used to view the Internet Explorer history
MiTeC Internet history browser	Allows you to see the history of all browsers visually
IECacheView	Allows you to gather information about Internet Explorer cache
IECookiesView	Allows you to gather information about Internet explorer cookies
ESEDatabaseView	It helps open the.dat file, which usually contains tables (login, cache, cookies) of the database
MozillaHistoryView	Shows the history of the Firefox browser
Magnet RAM capture	Allows you to dump live memory of the system
HashMyFiles	It calculates the hash of the evidence before and after the investigation
Encrypted disk detector	It is useful in analyzing the encrypted disk such as Bitlocker, TrueCrypt
Network miner	It is a network forensic analyzer that supports many protocols, detects operating system, hostname, packet-sniffing, and timestamp
SysInternals suite	Bundle of analyzing tools packed inside one suite

5 Network Forensics

Network forensics is a subdomain of forensics that helps to monitor and analyze the computer network to find malware activity and any possible intrusion in the organization. There are steps similar to computer forensics. However, we need a .pcap (packet capture) file to analyze, just like we had *.dmp* (dump) or *.mem* (memory) file in computer forensics.

The necessary steps are presented below:

Identification—the analyst will try to find an illicit incident based on the network indicators in the network traffic.

Storage and Capturing step signifies collecting or recording the ongoing network traffic using standardized software such as Wireshark, tcpdump, or tshark capture.

Analysis and Examination step signifies in-depth knowledge of OSI layers to extract useful information from any network dump. It reconstructs the network packet and draws relevant conclusions.

Incident Response is based on the analysis performed. An alert is triggered if any intrusion is detected, and an incident response team should tackle the attack.

In Table 9 are presented the commonly used network forensics tools along with their description.

There are many challenges while doing network forensics. The first challenge is to sniff the desired network traffic as sometimes it is difficult to sniff the network

Table 9 Network forensic tools—including sniffers and deep packet inspection (DPI)

Network forensics tool	Description
Wireshark	It performs network forensic to visually see the network packets and the content inside
Tcpdump	$ tcpdump –v forensicsdump.pcap
Tshark	$ tshark –i wlan0
Bro	Deep Inspection Packet Analyzer that allows deep packet inspection
Snort, Suricata	Network intrusion detection system
Xplico	Web user interface that reconstructs the content of the packet
ssldump	It extracts the SSL information from the dump packet

due to severe network configuration, security measures, and policies. The second major challenge is drawing a conclusion from unknown network traffic, which was not intended to be analyzed.

We can capture traffic from the target network in six different ways on a switched network as follows: port mirroring, ARP cache poisoning, flooding, DHCP redirection, and using a tap (with a hub). There are two working modes for a Network Interface Card (NIC): promiscuous mode and non-promiscuous mode. In the promiscuous mode, the NIC can receive all the data it can see, whereas in non-promiscuous mode, the NIC can see only its destinated traffic. By default, the NIC is configured to work in non-promiscuous, so it will drop all the packets which are not addressed to it. We can set the mode of NIC to promiscuous using the aircrack-ng tool, and once it is set to promiscuous, it can capture all data from layer 2 to layer 7 of OSI. For Windows, the switch to promiscuous mode should be done using WinPcap—Window Packet Capture—https://www.winpcap.org/, and for Linux, by using LibPcap—promiscuous capture library—https://www.tcpdump.org/.

Hubs are the best device when we want to sniff any network. Unfortunately, criminals always reconnaissance the network infrastructure that poses hubs in the infrastructure. With the help of the hub, we can easily capture the traffic from all the systems that are connected to the hub. In contrast to that, switches are intelligent devices that segment the traffic by observing the source and destination MAC addresses of every data frame that passes through it. The switch has a Content Addressable Memory (CAM) table, and when an Ethernet frame comes to the switch port, it observes the frame's source and destination MAC addresses. The source MAC address and the physical port of the switch are memorized in the CAM table. Then, the incoming frame is forwarded to that specific MAC address to the physical port that matched the address in the CAM table. If no match were found, the frame would be forwarded to all the switch ports, except the source one. The functioning of the switch is presented in Fig. 22. Generally, the switch works at the data link layer of the OSI model, but modern switches can work with higher layers too.

To capture any network traffic, it should have the following prerequisites:

- The device of which you want to sniff the traffic must be on your local network, or it could be an intermediary point.

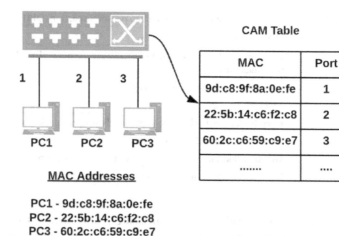

CAM Table

MAC	Port
9d:c8:9f:8a:0e:fe	1
22:5b:14:c6:f2:c8	2
60:2c:c6:59:c9:e7	3
....	....

MAC Addresses

PC1 - 9d:c8:9f:8a:0e:fe
PC2 - 22:5b:14:c6:f2:c8
PC3 - 60:2c:c6:59:c9:e7

Fig. 22 Working of a switch. The CAM table contains MAC and port numbers. Each frame is analyzed, the destination address is extracted and searched in the CAM table. If any match the frame forwarded to the destination, else the frame is sent to all the ports except the source one

- The device must be connected to a hub, switch, or border router through which traffic passes.

It is relatively easy to capture traffic from a hub: Just plug in our sniffer to the hub and start sniffing. The process is known as "Passive Sniffing", where there is no network modification nor infrastructure configuration. Whereas the scenario gets complicated when dealing with switches, it is hard to view traffic from other clients in the network as the switches segment the traffic. Therefore, we need to make extra efforts to get that traffic in the so-called Active Sniffing.

In Fig. 23, PC3 is a sniffer that wants to see all the traffic on the network where it is connected to. PC3 is connected to the switch, which provides segmented traffic using the CAM table, so PC3 cannot view any traffic of hosts 1 and 2 (PC1 and PC2). In order to see all traffic from the switch, there are several mechanisms we can apply, as presented in the following sections of this chapter.

5.1 ARP Cache Poisoning

Address resolution protocol (ARP) is a communication protocol that resolves a known IP address to an unknown MAC address of any host. This ARP request–reply is stored inside the ARP cache on each client for a short period. The cache stores IP address, MAC address, and Time to Live (TTL) for each entry.

When dealing with ARP in Wireshark, you will see that TTL entries get different for Windows and Linux-based systems. For example, it is 2 min in Windows systems, but for Linux, TTL is 15 min. So just by looking at the TTL of the ARP packets in a

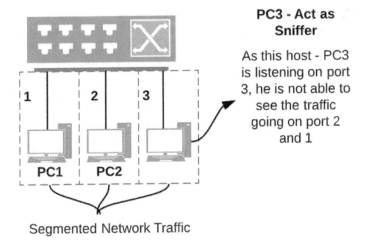

PC3 - Act as Sniffer

As this host - PC3 is listening on port 3, he is not able to see the traffic going on port 2 and 1

Segmented Network Traffic

Fig. 23 Concept of segmented network traffic adopted by the switch. As a sniffer, it is difficult to capture and sniff the traffic of other ports

network traffic capture, you can decide if the traffic is from a Windows-based system or from Linux one.

ARP is a relatively simple protocol and therefore gets easily exploited. Its functioning is presented in Fig. 24.

Our systems get easily manipulated to trust bogus ARP requests and replies from not correct or malicious devices. There are no certain ways to verify the ARP request, and there is no authentication mechanism for the responding device in order to make sure that it is really who it says it is.

ARP cache poisoning or spoofing is an attack where an attacker sends spoofed ARP requests to the hosts in the network. The legitimate hosts think that the request is from trusted hosts (it is not) and accept it, and later all the traffic directed between legitimate hosts will be diverted to the attacker system, which can inject any malicious payload in the packet as shown in Fig. 25. As ARP request cannot be verified, anyone can spoof it and make a bogus entry in the CAM table of the switch. This attack is a man-in-the-middle attack that can be performed using tools such as— Cain and *Abel* (https://en.wikipedia.org/wiki/Cain_and_Abel_(software)), *arpspoof* (https://github.com/ickerwx/arpspoof) and *Ettercap* (https://www.ettercap-project. org/). However, the attack can be detected by specialized software, like Wireshark, as shown in Fig. 26.

5.2 Port Mirroring

Switch segments the traffic using physical port number and MAC address, a sniffer who wants to view all traffic from the surroundings has to mirror the port, to which he

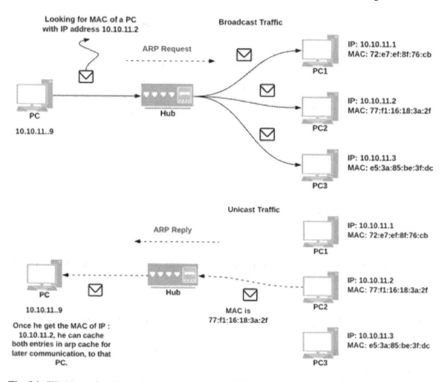

Fig. 24 Working of Address resolutionr protocol (ARP) with its request and reply packets. The ARP request is broadcast traffic. The destination will send an ARP reply (containing MAC) as Unicast traffic

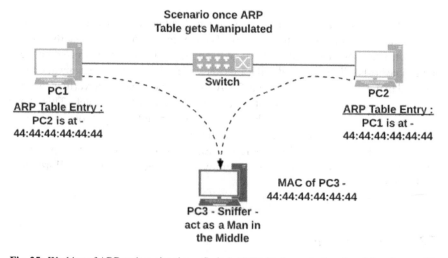

Fig. 25 Working of ARP cache poisoning—Switch ARP table is manipulated, and therefore, traffic between PC1 and PC2 is now directed through PC3

```
ARP        60 192.168.168.2 is at 00:0c:29:f7:35:b3
ARP        60 192.168.168.130 is at 00:0c:29:f7:35:b3 (duplicate use of 192.168.168.2 detected!)
ARP        60 Who has 192.168.168.130? Tell 192.168.168.2 (duplicate use of 192.168.168.2 detected!)
ARP        42 192.168.168.130 is at 00:0c:29:b9:d9:a0 (duplicate use of 192.168.168.2 detected!)
ARP        60 192.168.168.2 is at 00:0c:29:f7:35:b3
ARP        60 192.168.168.130 is at 00:0c:29:f7:35:b3 (duplicate use of 192.168.168.2 detected!)
ARP        60 192.168.168.2 is at 00:0c:29:f7:35:b3
ARP        60 192.168.168.130 is at 00:0c:29:f7:35:b3 (duplicate use of 192.168.168.2 detected!)
ARP        42 Who has 192.168.168.254? Tell 192.168.168.130
ARP        60 192.168.168.254 is at 00:50:56:f4:b3:7c
ARP        60 192.168.168.130 is at 00:0c:29:f7:35:b3 (duplicate use of 192.168.168.2 detected!)
ARP        60 192.168.168.2 is at 00:0c:29:f7:35:b3
ARP        60 192.168.168.130 is at 00:0c:29:f7:35:b3 (duplicate use of 192.168.168.2 detected!)
ARP        60 192.168.168.2 is at 00:0c:29:f7:35:b3
```

Fig. 26 Wireshark has detected an ARP spoof attack, where multiple entries of the same MAC address can be observed with different IP addresses

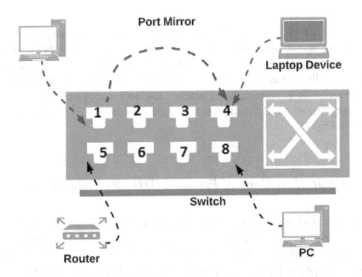

Fig. 27 Port mirroring—Port 1 is mirrored on port 4. Therefore, the laptop device can perceive the packets of PC1 on port 4, as it is mirrored

wants to sniff. In port mirroring, a copy of the sent packet from one switch port can be sent to a sniffer that is connected to another port (mirrored port). This process is known as SPAN—Switch Port Analyzer, and it is explained in Fig. 27. For example, PC1 is connected to port 1, and another device can mirror port 1 to another port (i.e., port 4). All data that comes to port 1 can now be copied/mirrored to port 4 and straightforwardly analyzed/visualized by that device.

5.3 Flooding

At earlier times, the conventional switches, usually cheaper and lower end, are not very secure. The switch has the CAM table, which is just a list of entries that an

19 0.001810730	107.150.59.123	192.168.168.132	IPv4	60
20 0.001922396	0.206.99.53	192.168.168.132	IPv4	54
21 0.001980060	110.24.125.104	192.168.168.132	IPv4	54
22 0.002103412	110.24.125.104	192.168.168.132	IPv4	60
23 0.002195586	220.142.7.38	192.168.168.132	IPv4	54
24 0.002254576	254.237.126.119	192.168.168.132	IPv4	54
25 0.002380199	220.142.7.38	192.168.168.132	IPv4	60
26 0.002389187	254.237.126.119	192.168.168.132	IPv4	60
27 0.002457567	78.62.91.16	192.168.168.132	IPv4	54

Fig. 28 Manipulation of switch CAM table using flood—macof tool generates a large number of flood packets, not bearable by the switch. It will start flooding out on all ports, so the switch will act as a hub

attacker can easily manipulate. The idea of the attack is to disintegrate the CAM table by filling up the table with multiple entries, so that switch cannot hold any more entries and hence it will start flooding out to all ports of the switch (act as a hub). That will allow an attacker to view the network traffic of others, which otherwise was not possible. To generate a flood attack, a tool like macof from dsniff—(https://www.monkey.org/~dugsong/dsniff/) can be used. The effect of a successful attack is presented in Fig. 28.

5.4 Dynamic Host Control Protocol (DHCP) Redirection

A cyberattacker can place a rogue Dynamic Host Control Protocol (DHCP) server instead of a legitimate DHCP server. The attacker will broadcast forged DHCP requests (using social engineering) and attempt to lease all available DHCP servers. As a result, the user will not get or cannot renew the IP address from the legitimate DHCP server. The attacker will grab this opportunity, and he will start his own DHCP server (rogued). The user's computer will access it to get to the Internet. However, the user's traffic doesn't go directly to the Internet but redirects first to the owner of the rogued DHCP server (attacker), which can now view the user's traffic (http://gobbler.sourceforge.net/), (http://dhcpstarv.sourceforge.net/).

5.5 Detection of TOR Traffic in the Captured Traffic

The analysis of the network traffic could lead to various important details. Millions of users are attracted to privacy browsers such as TOR, Freenet, and I2P. These networks are entirely encrypted and very difficult to examine; however, we can find bits and pieces of the information related to such networks in the evidence file. For example, a user who is attempting to open a website in the TOR network can easily be seen in the Wireshark TCP stream.

```
[Header checksum status: Unverified]
Source: Nilesh (192.168.43.52)
Destination: 66-206-0-82.static.hvvc.us (66.206.0.82)
Transmission Control Protocol, Src Port: 6010, Dst Port: 9001, Seq: 544, Ack: 1, Len: 543
   Source Port: 6010
   Destination Port: 9001
   [Stream index: 1]
   [TCP Segment Len: 543]
   Sequence number: 544     (relative sequence number)
   Sequence number (raw): 1615430267
O..........+.p,y.....qo>...TN...no.n......K/z.J          A.....1..-...=
D.q........'z.;.MU(.......Y.V..nk.i.q....&...A..G...A^.7.5.......9.....
...|1Y....h.Mk3..@Tor.."B.s+4(....X...H...U..)..}.+.@w(10.v@,...T.
```

Packet 2098. 411 client pkts, 1,674 server pkts, 591 turns. Click to select.

| Entire conversation (2457 kB) | ⌄ | Show and save data as | ASCII |

Find: tor

Fig. 29 Wireshark capture—captured TOR guard node—66.206.0.82. We can find the country from where the suspect has committed the crime. Whois can give the IP country location (in our case, IP belongs to—the USA, Tampa, Hivelocity Ventures Corp.)

In Fig. 29, we can see the TOR guard node with an IP: 66.206.0.82, which will remain the same for a TOR session. Also, there are exit nodes lists that are publicly available—(https://www.dan.me.uk/tornodes), but the most difficult part is to find the TOR middle node. Indeed, there are various ways to attack the TOR network [10].

To confirm our check on the TOR network, we can use a DPI network analyzer which will inspect packets to a deeper level to understand the behavior of user activity.

Bro or Zeek is an open-source network security monitoring tool that is based on DPI and can log various entities such as HTTP, SSL, ×509 certificates, FTP, and many others. From those logs, which have these unknown URLs shown in Fig. 30, we get to know and confirm that the user was using the TOR network.

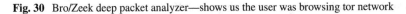

```
CN=www.mvhyr3o2mkvxiabipt.net
rsaEncryption     sha256WithRSAEncryption rsa      2048

1592961242.401103          Fmrxhq39YU3GCLX7ke        3
CN=www.qx7t57orw26t5u3azizu.com 1592895600.000000
65537
```

Fig. 30 Bro/Zeek deep packet analyzer—shows us the user was browsing tor network

6   Conclusion

Forensics rely on the investigator's skills and the evidence he caught from the crime scene. Uncovering a piece of information from an asset is a critical and challenging task due to several risks such as evidence perishability, inconsistent data, or no correlation between the information and the task cost. The investigator's skills come in handy to find the best possible solution to uncover the hidden tracks. In conclusion, the chapter deals with an introduction to cyberforensics and its phases. Furthermore, it contributes to showing how to gather valuable information from generally ignored locations, such as Windows registry, browser artifacts and Event Viewer logs. Several memory analyses and dumping tools have been introduced to dump the RAM from the system to find useful information. In addition to that, a short formal introduction to network forensics is presented, enabling the reader to learn and understand how to intercept and analyze traffic between stations. As future work, there is a real need to consider mobile forensics. Cybercrimes have significantly increased in mobile platforms; therefore, a formal introduction of forensics and insight learning is necessary to understand the peculiarities of this challenging domain.

References

1. S. Ellis, Cyber Forensics,in *Computer and Information Security Handbook* 573–602 (2013)
2. N. Galov, 39 Jaw-Dropping DDoS Statistics to Keep in Mind for 2021, in *Hosting Tribunal* (2021). https://cutt.ly/Ex8bame
3. R. Ruffolo, Fear, greed, lust: Phishing's sure-fire lures | IT World Canada News, in *IT World Canada—Information Technology news on products, Services and Issues for CIOs, IT Managers and Network Admins* (2021). https://cutt.ly/2x8mGpb. Accessed 30 March 2021
4. Z. Al-Sharif, M. Al-Saleh, L. Alawneh et al., Live forensics of software attacks on cyber–physical systems. Futur. Gener. Comput. Syst. **108**, 1217–1229 (2020)
5. C. Ngejane, J. Eloff, T. Sefara, V. Marivate, Digital forensics supported by machine learning for the detection of online sexual predatory chats. Forensic Sci. Int. Digital Invest. **36**, 301109 (2021)
6. L. Abrams, K. Andreou, T. Bradley, et al., Having fun with sysinternals. Winternals 441–457 (2006).https://doi.org/10.1016/b978-159749079-5/50015-9
7. R. Padmanabhan, K. Lobo, M. Ghelani, D. Sujan and M. Shirole, Comparative analysis of commercial and open-source mobile device forensic tools, in 2016 Ninth International Conference on Contemporary Computing (IC3), Noida, India, 2016, pp. 1–6
8. R. McDown, C. Varol, L. Carvajal, L. Chen, In-depth analysis of computer memory acquisition software for forensic purposes. J. Forensic Sci. **61**, S110–S116 (2015)
9. A. Bozkir, E. Tahillioglu, M. Aydos, I. Kara, Catch them alive: A malware detection approach through memory forensics, manifold learning and computer vision. Comput. Secur. **103**, 102166
10. J. Salo, Recent attacks on Tor, in: Aalto University (2010), Finland

Chapter 6
Intrusion Detection Systems Fundamentals

1 Introduction to Intrusion Detection System

There is a constant increase in cyberattacks in the last several years. Ransomware, denial of service (DoS), phishing, and unauthorized access are some of the major threats to any organization responsible for providing digital services to the users. The malware infection is one of the brutal attacks, and its severity greatly loosens our line of defense. Infection growth is exponential from 12.4 mn in 2009 to 812 mn in 2018 [1]. If we analyze recent cyberthreats of 2020, then on average, 81% of vulnerabilities are network-based and the rest of them are application layer-specific vulnerabilities. In the stack of attack, malware is having highest severity and releases its competent power when they come in contact with any network; it attacks largely and simultaneously spreading from subnet to subnet. Recent attacks show that every 4 s, a new piece of malware is being evolved and try to penetrate the system with malicious intentions [2]. Companies that provide services such as financial, medical, digital, or any telecom service are the focal point of the attackers, as such companies are having a tremendous amount of information that can be stolen and sold out in the black market, such as TOR. The growing number of cyberthreats is our concern; however, our main distress is understanding the attack mechanisms. The attackers keep gaining knowledge and prepare an attack with more severity than previous, and in such cases, we do not get any time to stop or prevent it [3, 4]. Figure 1 clearly shows us the breaches of 2018 in the USA. According to reference, the highest number of attacks are access-related breaches, and 20% of these breaches are from employee e-mail [5].

The blue team, which is our line of defense, uses several preventive measures ranging from firewalls and sophisticated antivirus. Unfortunately, these traditional methodologies are inefficient in protecting against attacks. More than that, not even the complex protection methods such as hardening operating systems is not successful. The growing rate of newer technologies connected to different complex networks, storage infrastructure, or online platforms such as the cloud becomes more complicated to understand and handle. With such technologies, it is hard to find and

© The Author(s), under exclusive license to Springer Nature Singapore Pte Ltd. 2022
N. Dutta et al., *Cyber Security: Issues and Current Trends*, Studies in Computational Intelligence 995, https://doi.org/10.1007/978-981-16-6597-4_6

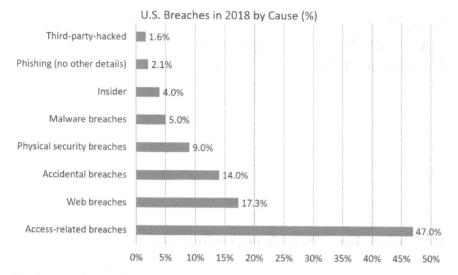

Fig. 1 Topmost breaches in USA in 2018. The US State Attorneys General published these data breaches in 2018 showed that access-related violations constituted the bulk of successful breaches. *Source* www.f5.com/labs

remove the vulnerabilities. As a result, we are giving a large scope and a scalable environment (electronic world) to the attackers, where they can come and play their wicked game of hacking. For them, the ground is too large to play, but for cybersecurity specialists, it is not, as they cannot be everywhere to protect the system. To tackle such a situation, there is a constant need for a newer technology that protects our system and alerts the system administrator before an attack occurs. For that reason, IDS/IPS has evolved and acts as a system that detects the attack and prevents it before happening.

2   Techniques to Combat Cyberthreats

Numerous techniques are already available in the market to fight these threats, but they are not reliable and flexible to put in our protective measures. Moreover, individually, these protective measures cannot go that far to protect our system. Hence, we need some heuristic or hybrid approaches that can easily disclose the nature of an attack. This section will discuss some of the protective measures, emphasizing their advantages and disadvantages.

2.1 Firewall

A firewall dramatically helps in the network security system since many applications and services want to connect to the network (public or private). A firewall can allow or deny traffic; it controls and monitors incoming and outgoing network traffic based on the rule that was defined in the ruleset. Thus, it provides an excellent defense perimeter around the organization to protect it from external threats as well as from internal threats. Organizations integrate firewalls in their security information and event management (SIEM) systems along with other security devices.

They are categorized into two types—network firewall and host firewall. Network firewall deals with the local network to control the traffic between the protected network and other networks. The host firewall is installed on a physical or virtual machine, and it will monitor the traffic that is going in and out from that machine. It has the potential to inspect packets transmitted between the computer and the Internet or between computers, based on the packet filter's rule. If any packet violates the rules, then it is simply dropped (discarded); otherwise, the packet is allowed to pass.

The next-generation firewall (NGFW) can work on the application layer of the Open System Interconnection (OSI) model. This is because it understands the application and its underlying protocols, such as File Transfer Protocol (FTP), Hypertext Transfer Protocol (HTTP), Domain Name System (DNS), and Simple Mail Transfer Protocol (SMTP). Therefore, it will deny any application request that does not satisfy the firewall rule, the protocol that is not listed and runs on an allowed port.

Packet Filtering Firewall—These firewalls will filter the attributes such as source IP address, destination IP, source port, and destination port number. The decision of forwarding or discarding the packet is based on the rule set defined in the firewall. For instance, if the rule is set to block all FTP connections, listening on port 21, then any packet that is destined to transmission control protocol (TCP) port 21 would be dropped. However, too much nuisance happens in these types of firewalls, as it never analyzes protocols, and moreover, it is vulnerable to IP spoofing attacks [6].

Stateful Filtering Firewall—They are an on-go filtering system that will monitor the packets running between endpoints and maintain a table of previous connections. Later, these connections can be used for comparing whenever any new packet comes in. Based on the comparison result, the firewall decides if the incoming connection is a part of a valid connection from the table or not. Hence, it can prevent several DoS and spoofing attacks. In reference [7], the authors had checked the credibility of stateful firewalls in a high-level declarative language.

Proxy Firewall—This type of firewall is settled in the application layer, where a packet and stateless filtering is not helpful. In the client–server model, the server manages all the requests from the clients and sends responses according to the used applications and protocols. In order to protect the server from any external threat, organizations use one more layer of security, where instead of using the real IP address, they use a proxy. In this way, the request is redirected from proxy to the intended user without disclosing the actual IP address. The incoming request packet is compared with the firewall rule, and then and only then the proxy firewall opens the connection to the requested server. This will block many attacks such as malware specific, DoS, unauthorized access, and file execution attacks.

2.2   Authentication

Authentication is the process of verifying the user's identity to access any specific system. It will help in validating the user's credential (document, login credential, etc.) in order to ensure that any external entity doesn't counterfeit; here, user will prove himself he is who he says he is by providing valid credential for authentication and later for authorization to use any service. The user will generally show his username and password for authentication, but sometimes, he also provides other elements known as factors. These factors can be combining with his login credential, and the authentication process can be classified as follows:

(1)   *Single Factor Authentication*—User will only provide a username and password to get authenticated to use any website.

(2)   *Two Factor Authentication* is a combination of login credentials and something that could be a "PIN," a code, or a biometric feature.

(3)   *Multifactor Authentication*—It takes more than one factor to authenticate, such as username, password with PIN and with biometric factor—facial recognition or fingerprint.

(4)   *One-Time Password (OTP)*—This is an automatically generated alphanumeric character sequence, active for few minutes to use to authenticate. It is used for one login session, new users, or users who lost their password.

(5)   *API Authentication*—It gives an application the ability to communicate with the API server for authentication purposes.

    a.   HTTP Authentication—It requests the user to provide a username and a password to access the system. The request mechanism consists of the server sending a 401 header (unauthorized access), which forces the client to authenticate. Subsequently, the client will send an authorization request header with its credentials.

    b.   API Key Authentication—Many third-party services that have to be used in a website use such as Google Webmaster, Google Maps, or Google Analytics, require an API key for authentication. The key is generated by the third-party system and should be introduced on the website. Then, each time when the website is accessed, the third-party service will identify the website using the API key, which was pre-registered.

    c.   Open Authorization (OAuth)—It is mainly used to authenticate yourself using third-party services. For example, you want to log in to some website, which has a login control panel. You can log in using the same password by which you have registered. You can also log in with "Login

with Facebook" or "Login with Google." OAuth can act as an intermediary service that uses a token-based authentication system, where users
try to log in with the server using a token (temporary credential request),
as shown in Fig. 2.

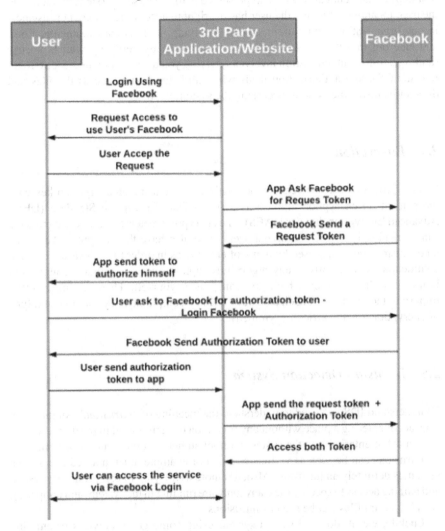

Fig. 2 Sequence diagram of Open Authorization (OAuth). Sequence operation between user, application, and Facebook for Open Authorization (OAuth)

2.3   Authorization

Many systems use a two-step process to authenticate and then authorize the users. The first step is to authenticate the user, as presented before. The second step is to authorize the user, meaning validating the user has the right to access a resource. For instance, in any organization, you can enter it by verifying your ID-card (authentication), and entering into the server room, you can enter it by verifying your biometric (authorization). Authentication involves verifying you to enter into the system (login into the OS), and authorization deals with validating you to access the files and directories inside the system (accessing the system files).

2.4   Encryption

It is the process where one can encode their message using an encryption key with the help of an encryption algorithm such as—Data Encryption Standard (DES), Advanced Encryption Standard (AES). The encryption means that the coded message can be read only by the authorized users, who also have the decryption key. This mechanism is generally used by many of enterprises in order to achieve secrecy and confidentiality. Currently, many algorithms support encryption, but the algorithm is not only the entity by which you can protect yourself. There are many other important factors as well, such as—key size, key complexity, key rotation, and type of encryption—symmetric or asymmetric.

2.5   Intrusion Detection System

To understand IDS, we need to understand the meaning of "*intrusion*." An intrusion is an act of entering a place without any invitation or permission. In the digital world, it is an act of entering into the system without authentication or authorization.

Intrusions can be seen as various events. For example, unauthorized access to a system is definitely an intrusion. Also, a remote user trying to compromise a system and gaining access to specific files, any malware running in the system, and tampering with important files can be seen as intrusions.

Luckily, we all do is to scan a system using some good antivirus or entering rigorous firewall configuration, but that is not enough to protect our system, as the signature of the malware can be rewritten, and any antivirus will not detect it. Also, the firewall rules can be revamped by the intruder.

To achieve maximum protection, we need to combine these protective measures, but it is not an easy task as it is hard to configure each and every entity which understands and operates differently. Perhaps, we can use a platform that provides all

Fig. 3 Statistics of intrusion detection system market. IDS/IPS market size in post-COVID-19 is getting higher every year due to increased number of intrusions and various policy violations. *Source* www.marketsandmarkets.com

these functionalities in one single software, and hence, we need an intrusion detection system—a monitoring system that could be a hardware platform or a software application that continuously observes the incoming and outgoing packets from the organization. Any intrusion, malicious packet, or policy violation could be reported by the IDS, recorded in log files, and simultaneously send an alert to the system administrator. This technology is offered to security vendors or any independent user who want to protect their resources. The global market of IDS will grow from USD 4.8 billion in 2020 to USD 6.2 billion in 2025 and a compound annual growth rate (CAGR) to 5.4% from 2020 to 2025 [8], as shown in Fig. 3.

Recent trends on IDS also involve machine learning algorithms to enhance the power of monitoring. These algorithms are better at predicting future attacks and classify legitimate network traffic.

There are three different categories of IDS, which will be presented in the following subsections of this chapter:

(1) Network-based intrusion detection system (NIDS)
(2) Host-based intrusion detection system (HIDS)
(3) Distributed intrusion detection system (DIDS).

3 Network-Based Intrusion Detection System (NIDS)

A network-based IDS is used to observe and analyze the entire segment or any specific network subnet. It helps organizations protect their cloud, on-premise infrastructure, and hybrid environment for any malicious event indicating any compromise. NIDS

Table 1 Various network-based intrusion detection system

Product	Description
Snort	It is an open-source network-based intrusion detection system; it possesses the capability to observe real-time traffic and packet logging. https://www.snort.org/
Bro (Zeek—open-source network security monitoring tool)	It is a deep packet inspection analyzer, acts as NIDS. It operates at application and hence supports a large number of protocols. https://zeek.org/
Security onion	It is a packed Ubuntu configured with various network security tools used for analysis and monitoring purposes. https://securityonion.net/
Wireshark (earlier ethereal)	It is an open-source packet sniffer and packet analyzer. https://www.wireshark.org/

rigorously checks any policy violation, port scanning, illegitimate source, and destination traffic. This system is "passive" in nature, as it scans and alerts on finding any suspicious traffic to the network administrator. Therefore, it is deployed together with intrusion prevention systems (IPS), which are "active" in nature (Table 1) shows the various open source network-based intrusion detection system.

To work with NIDS, we need to change Network Interface Card (NIC) functioning mode, which generally utilizes "non-promiscuous mode." In this mode, the NIC listens only for the packets which are destined for its MAC address, and other packets are ignored. We need to change the mode to "promiscuous mode," when the NIC listens and accepts all the packets within the network. We can configure our NIC with promiscuous mode to sniff and send all packets through the NIDS to the destination system. This NIDS (act as a mediator entity) can observe and analyze the traffic between source and destination. The advantage of this category of IDS is that it doesn't add any load to the system, and on top of that, the attacker never knows that they are monitored continuously and hence NIDS can't be touched by an attacker. Placing NIDS in a specific location and configuring it for proper functioning is the major concern of any network administrator or analyst. In Fig. 4, if any server is compromised due to any weak configuration of NIDS, it leads to exploitation, where those servers could become hotspots to exploit other sever in the system [9].

4 Host-Based Intrusion Detection System (HIDS)

There is a slight difference between NIDS and HIDS, but both IDS types are effective in our systems. HIDS can protect the system or host on which it is installed and not the entire subnet. Here, the NIC can work in non-promiscuous mode to tackle any detection. It monitors file size and checksum that ensures the integrity of the file is maintained. HIDS could detect even a slight increase in the file size. It will intercept

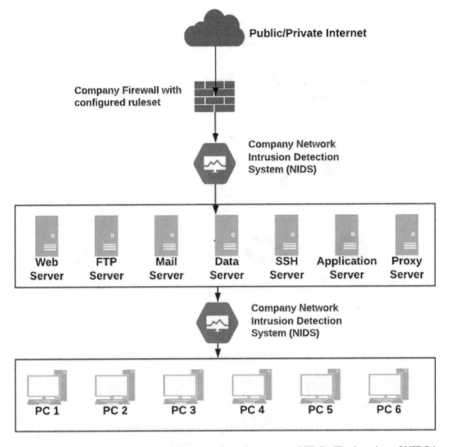

Fig. 4 Installation of network-based intrusion detection system (NIDS). The location of NIDS in the organization is important, as it can either be placed internally or externally near the company's firewall

any fake call that exploits a vulnerability in the system by continuously listening to the traffic within the system; furthermore, it never leaves the network.

In contrast to NIDS, HIDS adds load to the system, as it uses CPU resources from the system on which it is installed. The authors of reference [10] present a survey on standalone HIDS and current specific research trends. They discuss intrusion detection, file integrity, traffic analysis, behavior monitoring, and countermeasures of tampering in HIDS.

In Fig. 5, we can see that HIDS should be appropriately configured on each host system which you need to protect. For instance, if the server is an e-mail server, then your HIDS should be configured according to the e-mail server activity (rules are defined according to the e-mail activity), similarly for the web server and others. Some of the best HIDS available in the market are—open-source host-based intrusion detection system (OSSEC)—(https://www.ossec.net/), Samhain (https://la-samhna. de/samhain/) and security onion.

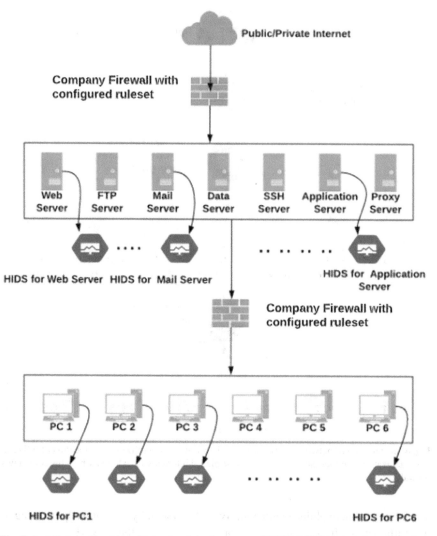

Fig. 5 Installation of host-based intrusion detection system (HIDS). HIDS is a host-based interface program. Therefore, the system must be installed to be load balanced, as HIDS uses system CPU

5   Distributed Intrusion Detection System (DIDS)

Distributed intrusion detection systems are sensor-based IDS, which combine both advantages and characteristics of NIDS and HIDS, as described in the previous sections. We can place both NIDS and HIDS sensors all over the organization to which you're preventing intruders. These sensors gather information about log files, error reports, store them, and later use them for analysis at a central location—server station, as shown in Fig. 6. DIDS is capable of packet sniffing, log analysis, malicious

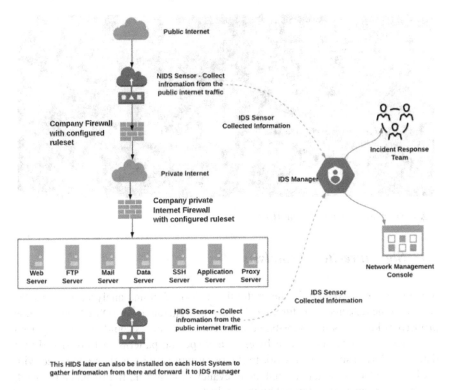

Fig. 6 Installation of distributed intrusion detection system (DIDS)

interruption, and file system analysis (sudden change in file size). The tool for DIDS depends on what system you want protection, such as if you're going to protect the cloud, then the sensor you wish to is meant explicitly for cloud protection. Authors of reference [11] designed and implemented a prototype DIDS that combines both distributed monitoring and data reduction. All IDS information is forwarded to a centralized system for further analysis. Furthermore, DIDS has a loophole in its centralized system because if this system is breached, the attacker can grab a large amount of information about the networks, their subnets, and confidential enterprise data.

IDS can further be classified based on their detection method as follows:

(1)   Signature-based analysis
(2)   Protocol-based analysis
(3)   Anomaly-based analysis (Fig. 7).

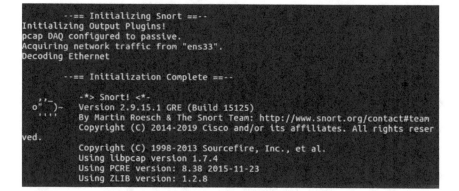

Fig. 7 Snort installation in Ubuntu 16.04

5.1 Signature-Based Analysis

Signatures are the pattern in the traffic that is looked at and analyzed by the IDS. Based on the detected signature, an alert can be generated. An IDS which uses this property is known as signature-based IDS. These signatures could be anything from "0's" and "1's" or a number of bytes or any specific patterns that are stored in a database. Any suspicious network traffic or system activity is then compared with the known attack signature to find the relevancy of an attack. If any match is found, an alarm is generated and noted by the analyst.

```
alert tcp any any -> any any (msg: "FTP Connection Alert" ;
                     sid:1000002)
```

This rule will alert any FTP connection to the network IP address it is destined to in your network. The meaning of "any-any" in the rule is the source and destination, and *sid* is the signature id, more complex rules we will cover in a later section. This analysis has some weaknesses as signatures are nothing but a pattern stored inside the database that proficient malware authors or network specialists could modify. Moreover, the known attack signature database doesn't need to have all signatures; with new capabilities in digital technologies, malware authors are writing complex malware that can pass this detection method easily.

5.2 Anomaly-Based Analysis

Due to certain loopholes in the signature-based analysis, there is a need for another approach to tackle this situation, namely anomaly-based analysis. If there is an attack and its signature or behavior are unknown to the IDS, we can apply the anomaly-based

Table 2 True and false alarm of intrusion detection system

Observed parameter in detection system	Description
True positive rate	Attack detected and alarm raised
True negative rate	Attack not detected and no alarm raised
False positive rate	Attack not detected, but alarm raised
False negative rate	Attack detected, but no alarm was raised

analysis to detect abnormalities inside the network. Using this approach, you can detect not only unknown attacks but also detect well-known attacks. The classification of legitimate and illegitimate traffic is based on heuristic or specified rules rather than with signature patterns. This is a detection method, and therefore, we need an automated machine that can classify and recognize the attack. We can use a neural network or machine learning for this purpose, where our detection method goes in two phases—the training phase (where the method will understand the normal operation of the system) and the testing phase (where this method will compare normal traffic with malicious traffic) to detect any anomaly. Also, other detector technologies can be used as well, such as data mining, grammar methods, and artificial immune system. There are some representative tools that support anomaly detection, for example, Snort (we'll discuss in a later section), Suricata (https://suricata-ids.org/) or Bro (Zeek—open-source network security monitoring tool). The disadvantage of this method consists in the existence of false positive alarms, which will raise an alarm even though there is no actual attack, as shown in Table 2. Reference [12] shows the comparison between support vector machine and random forest to detect and reduce the false alarm rate in IDS.

6 Snort—Network-Based Intrusion Detection System

Snort is one of the best-suited IDS that any company or an individual can deploy as a protective measure to secure their systems. It is an open-source tool, introduced in 1998 by Martin Roesch (CTO of Sourcefire), and later in 2013, it was purchased by Cisco [13]. Snort is on version 3.0 (beta) and works with all major operating systems at the time of writing. It acts as IDS as well as intrusion prevention system (IPS), which works on both signature and anomaly-based analysis; moreover, it is a packet sniffer and packet logger. It observes real-time network traffic and detects any suspicious intrusion with the help of its ruleset, which is continuously updated. For years, it is updating continuously to add other features. Also, add-on programs can be used to maintain log files, configure new signatures and rulesets, add new detection mechanisms (algorithms), and enhance alert mechanisms. However, these add-ons are third-party services and didn't come with Snort suite. Snort understands and works with TCP/IP, although we can arrange something in the suite that makes him support other protocols as well. Snort has many components, and cumulatively, these components make Snort competent in the IDS market.

A network packet is going to be captured by NIC, and it is sent to the decoder. By *"packet decoder"* we understand an entity that will try to find out the protocol inside the packet and compare the data against the behavior of the packet. For example, if the data is normal network traffic, it contains routine information. But if from the behavior aspect, it is a long string of data, its size is in GB or in TB, and then, possibly it is a "buffer overflow attack."

Furthermore, it will check the header of the protocol, the size of the header, and any abnormal TCP options. After this checking, it passes the packet to *"preprocessors."* They were introduced in the Snort version 1.5, which will parse the important information from incoming data packets from decoders. It will detect port scanning, reassemble the TCP fragments, find out any malicious intent in the stream of data, detect many types of attack, such as ARP spoofing, and it will inspect HTTP packets. Once the data has been passed from preprocessors, it is next passed to *"detection engine,"* where rules are configured. The rules are then compared with incoming data packets from decoders and preprocessors, and the rules in Snort are settled in the file—snort.conf. Finally, when the comparison between the data packets against the ruleset is completed, if any malicious activity is detected, the IDS will log in to the log files and alert the system administrator for further action. For that operation, Snort has another component, namely *"Alerting and Logging."*

6.1 Additional Snort Add-Ons

(1) *SGUIL (Snort GUI for Lamers/z)*—It provides the GUI base for Snort, consisting of server, sensor, and client, which will help IDS manager record the data for analysis and provide full packet capture. (https://sourceforge.net/ projects/sguil/).

(2) *OinkMaster*—It will help to update your Snort rules with current signatures and attack mechanisms. To download the rules from the various Snort packages, oinkmaster uses oinkcodes, which act as API keys associated with your user account. You can see oinkcodes only when you have a subscription for the ruleset when you get something a link similar to the following—https://www.snort.org/rules/snortrules-snapshot-2983.tar. gz?oinkcode=7299b976d49d48fd4d544714a21a011d172e6ba5.

(3) *Analysis Console for Intrusion Detection (ACID)*—It is another GUI-based platform for collecting and analyzing information to a web interface. To work with ACID, you need to install a web server (Apache), database server (MySQL), PHP, and other tools. It inspects source, destination addresses, and ports numbers, view different aspects of the packet (the size, header information, data, and so on). (http://acidlab.sourceforge.net/).

(4)   *Swatch*—It is a watcher which can see the log files more efficiently. It is written in Perl languages which can generate alerts and create an automation process. (https://sourceforge.net/projects/swatch/).

(5)   *SnortSnarf*—The tool is written in Perl language, where it converts Snort data into web pages for better inspection. It will interpret the log files and extract the data from the database server which you had installed. (http://www.silico ndefense.com/software/snortsnarf/index.htm).

(6)   *IDS Policy Manager*—When dealing with DIDS, we need to have a sensor-based application, and as it is a wide platform inspection, we need to apply the updated rule and its policies constantly. This tool will guide us to manage our IDS sensors remotely, update and configure the rules. (https://ccm.net/dow nload/download-5936-ids-policy-manager).

(7)   *SnortPlot.php*—It is Perl-based script that will allow us to visualize any attacks detected by the Snort. However, it is an obsolete package now.

(8)   *PigSentry*—It allows us to get real-time Snort alerts, it maintains a record of all alerts from past and future in a table, and later, it will match with any incoming attack with this table to create an alert.

6.2   Installation of Snort in Linux

Snort is a flexible platform that can be installed and deployed on any major operating system. The machine on which you are installing has some system requirements—at least 4 GB of RAM, efficient CPU/GPU, larger storage device (1 TB hard disk). We install Snort on the virtual machine, based on Ubuntu Linux distribution (version 16.04) [14]. Before moving further, you need to upgrade and update your machine so that all packages are up to date and running the latest version. Snort installation depends heavily on various dependencies, for example, for secure communication—SSH server, packet sniffing—libpcap, and low-level network configuration—libdumbnet, different parsers—bison and flex, and many more. You can install all the packages at once with this one command.

```
# apt-get install openssh-server ethtool build-essential
libpcap-dev libpcre3-dev libdumbnet-dev bison flex zlib1g-
dev liblzma-dev openssl libssl-dev
```

Snort has added one more feature in its 2.9 version—data acquisition library (DAQ). Before that version, any packet logging or sniffing would be done directly by libpcap, but with DAQ, it replaces all those calls to libpcap function and becomes an abstraction layer that handles all packet I/O and provides a variety of operations to hardware and software. More details regarding DAQ can be read from here (http://manual-snort-org.s3-website-us-east-1.amazonaws.com/node7.html#SECTION00254000000000000000). Following steps are for downloading DAQ, installing and configuring it.

```
# wget https://www.snort.org/downloads/snort/daq-
2.0.7.tar.gz
# tar -xvzf  daq-2.0.7.tar.gz
# cd daq-2.0.7
# ./configure
# ./make
# ./make install
```

This will configure the data acquisition package. We can write./*configure make* and *make install* in one line as well—./*configure && make && make install*; however, if any error comes up, you can separate out the command, and it will work fine../*configure* command will generate a *makefile*, *make* is responsible for reading and executing the code in make file—it has a header—"*install*," later used for installation purpose. *Make install* is accountable for the final step, reads the information from the install section for make file, and distribute (.exe or.dmg) and other files to the proper location in the file directory system of that machine. All this information is presented in the documentation of the DAQ system.

The next step is to download the Snort source code. When you are installing it on Linux, it is recommended to install via building source code, as it is not always possible that Snort repositories are having the latest version of Snort, but the source code is always in the latest version.

```
# wget https://www.snort.org/downloads/snort/snort-
2.9.16.tar.gz
# tar -xvzf snort-2.9.16.tar.gz
# cd snort-2.9.16
# ./configure --enable-sourcefire && make && make install
```

If in the command./configure any error comes up (LuaJIT Library not found), then you should modify it by—./configure—enable-sourcefire—disable-open-appid; however, you can also install the library from here—https://luajit.org/install.html, the JIT is Just in Time Compiler for Lua—language for embedded application. Here— enable-sourcefire is doing nothing much, but building some option for sourcefire (it will work even if you don't write). Ldconfig—will create links and cache to the recent shared libraries specified in the command. In addition to that, we will also use the "*ln*" command to create a symbolic link (soft link) that will create a reference for another file for calling purposes. There are two types of links in Linux—symbolic link (soft link) and hard link. Soft link will provide a symbolic path (reference path) that indicates the abstract location of another file. (ln –s \.\.file1 \.\.\.\ file1) and hard links are the specific location of the physical data.

```
# ldconfig
# ln -s /usr/local/bin/snort   /usr/sbin/snort  - Soft Link
# Snort -V
```

Once Snort is appropriately installed, we need to configure it. If you see the snort.conf file in the Snort folder, there are many paths containing folders and files, which we have to create manually now, such as—rules, preprocess rules, white list rules, and black list rules, for that we will use mkdir and touch command to create folder and files.

```
# mkdir /etc/snort
# mkdir /etc/snort/preproc_rules
# mkdir /etc/snort/rules
# mkdir /var/log/snort
# mkdir /usr/local/lib/snort_dynamicrules
# touch /etc/snort/rules/white_list.rules
# touch /etc/snort/rules/black_list.rules
# touch /etc/snort/rules/local.rules
```

We will change the file permission of specific files using the Linux command utility—"*chmod.*"

```
# chmod -R 5775 /etc/snort/
# chmod -R 5775 /var/log/snort/
# chmod -R 5775 /usr/local/lib/snort
# chmod -R 5775 /usr/local/lib/snort_dynamicrules/
```

Here, 5775 is an octal value for file permission, which signifies the symbolic value for -rwsrwxr-t, -r(read) -w(write), -x(execute) -s (setuid) if is in the user, (setgid) if it is in the group, -t is used for sticky bits (flag bit). More can be found in the command reference, which is accessible https://chmodcommand.com/chmod-5775/.

Next, copy all the file that has the following extensions: *.conf *.map *.dtd to /etc./snort/. If you couldn't find any file with this extension in snort/etc. folder, then find it in other folders, we will copy it to the root directory of /etc./snort. Copy all the dynamic preprocessors and paste it in */usr/local/lib/snort_dynamicpreprocessor/*

```
# cd snort-2.9.16/etc
# cp -avr *.conf *.map *.dtd /etc/snort/
# cp -avr src/dynamic-preproce
sors/build/usr/local/lib/snort_dynamicpreprocessor/*
/usr/local/lib/snort_dynamicpreprocessor/
```

Move to /etc./*snort/snort.conf* for configuring specific attributes such as IP address and modifying the default paths of rules.

```
# gedit /etc/snort/snort.conf
```

It will open the snort.conf, and changes are as follows,
Setup the network addresses you are protecting

ipvar HOME_NET 192.128.128.10

Path to your rules files (this can be a relative path)
Note for Windows users: You are advised to make this an absolute path,
such as: c:\snort\rules

> var RULE_PATH /etc/snort/rules
> var SO_RULE_PATH /etc/snort/so_rules
> var PREPROC_RULE_PATH /etc/snort/preproc_rules
> var WHITE_LIST_PATH /etc/snort/rules
> var BLACK_LIST_PATH /etc/snort/rules

Lastly uncomment the following line, include $RULE_PATH/local.rules. Save it and exit. We need to evaluate and verify the changes we did are correctly modified or not.

```
# snort -T -i ens33 -c /etc/snort/snort.conf
```

In this command –i is for interface your Internet is connected (my interface is ens33, you're could be eth0—verify it using—ifconfig command), -T stands for test and report on current Snort configuration, -c specifies the rule files (snort.conf). If everything works properly, you will the successful message—"Snort successfully validated the configuration." To test the functionality of IDS, we will create some simple rules and check that our IDS detects them or not. To write your own rules, you can enter the path /etc./snort/rules/local.rules. Here are the simple rules that we jotted inside the file.

```
alert tcp any any -> $HOME_NET 21 (msg:"FTP connection at-
tempt"; sid:1000001; rev:1;)

alert icmp any any -> $HOME_NET any (msg:"ICMP connection at-
tempt"; sid:1000002; rev:1;)

alert tcp any any -> $HOME_NET 80 (msg:"TELNET connection at-
tempt"; sid:1000003; rev:1;)

alert tcp any any -> $HOME_NET 22 (msg:"SSH Connection Detect-
ed"; sid:103; rev:1;)
```

Later, we connect this machine on which IDS is installed with another machine via several connections such as FTP, Telnet, SSH as in the rules specified. When we try to connect this machine, the IDS is logging every connection in the backend and alerting simultaneously with the below command. In the command—A will print the output in the console, -q will not show other unnecessary information such as banner information (information related to the machine—name, version, running services on that machine, and so on), -c is the path of our configuration file—snort.conf, -i is the interface by which our machine is connected to the Internet.

```
# snort -A console -q -c /etc/snort/snort.conf -i ens33
```

You will see Snort will detect all this connection in the console window. We save that console window in the.txt file, and here is the preview of that file; furthermore, you can also see all this information in the log file too, which is stored in /var/log/snort/. You can read this file using tshark –r or tcpdump –r.

```
04/28-06:08:31.798621 [**] [1:1000002:1] ICMP connection at-
tempt   [**]   [Priority:   0]   {ICMP}   192.128.128.1   ->
192.128.128.10

04/28-06:08:42.676615 [**] [1:1000003:1] TELNET connection at-
tempt   [**]   [Priority:   0]   {TCP}   192.128.128.1:5639   ->
192.128.128.10:80

04/28-06:08:59.303763 [**] [1:103:1] SSH Connection Detected
[**]   [Priority:   0]   {TCP}   192.128.128.1:5658   ->
192.128.128.10:22
```

Snort will read the incoming packet in the form of a.pcap file, pcap is using processpacket() to read the packet, and later, it calls the decoder component of Snort to decode the packet at each layer. Each packet in Snort will move from one layer to the other layer of OSI using pointers. Hence, the pointer will point to one decoder to decode one specific layer and moves to the next decoder to decode another layer and so on. Decoder information is stored at /snort/src/decode.h. The decoder will pass this network packet after decoding to preprocessors and then from preprocessor to detection engine, and this phase consist of tagging—used to tag any packet, which hasn't triggered any rule. This is additional traffic that an analyst can analyze.

```
alert udp any any -> any 123 (msg:"Unknown Traffic"; flag:s;
tag:session; 10, seconds; sid:1000001; rev:1;)
```

This is the traffic for Network Time Protocol (NTP), tagged as a session log. The rule gets triggered whenever it starts its session on port 123 for the next 10 s. We can also use the feature *threshold for logging and alerting*, which will limit the number of logged alerts. The primary concern of the threshold is to reduce the number of false alarms. For example, to detect TOR usage in Snort, we can simply use the port number (9001, 9030) to restrict the network traffic belongs to TOR. We need to provide this port number to snort.conf and then write the below rule in local.rules.

```
alert tcp $HOME_NET any -> $EXTERNAL_NET 80,443,9001,9030
(msg:   "TOR   Usage   Detected;   classtype:policy-violation;
sid:50009;)
```

```
04/28-07:34:06.106403 [**] [1:50009:0] "TOR client access de-
tected" [**] [Classification: Potential Corporate Privacy Vio-
lation]   [Priority:   1]   {TCP}   192.168.168.129:41456   ->
217.182.75.181:9001

04/28-07:34:13.129683 [**] [1:50009:0] "TOR client access de-
tected" [**] [Classification: Potential Corporate Privacy Vio-
lation]   [Priority:   1]   {TCP}   192.168.168.129:38502   ->
185.97.32.34:9001

04/28-07:34:16.083041 [**] [1:50009:0] "TOR client access de-
tected" [**] [Classification: Potential Corporate Privacy Vio-
lation]   [Priority:   1]   {TCP}   192.168.168.129:41358   ->
163.172.149.155:443
```

6.3   Snort Rules

Snort rules are powerful enough to detect unknown network traffic. However, there are many attributes that you need to remember while creating the rules. These rules should be in one single line, as the Snort rule parser doesn't know to handle multiple lines. Snort rules are divided into two sections—rule header and rule option. The rule header contains rule action, protocol, source and destination IP address, netmasks, and port number.

```
alert tcp any any -> $HOME_NET 21 (msg:"FTP connection at-
tempt"; sid:1000001; rev:1;)
```

In the above example, till the first parenthesis, it is the "rule header," and after that, whatever in the parenthesis is the "rule option," as shown in Table 3. The first word of this rule—alert is the "rule action." There are many such rule actions described as,

Table 3 Snort rule header—rule action

Rules	Description
(Alert tcp ! 192.168.168.129.0/24 any → 192.168.168.130/24)	Match all IP addresses except (192.168.168.129/24) for inspection
(Alert tcp ! [192.168.168.129.0/24, 10. 10. 10. 1/24] any → [192.168.168.130/24, 10. 10. 10. 2/24])	Match all IP addresses except these IP addresses enclosed in [] for inspection
Log TCP any any → 192.168.168.130/24 1:1024	Log all TCP traffic for any IP address and port number to destination IP address (192.168.168.130/24), destination port ranging from 1 to 1024
Log TCP any any → 192.168.168.130/24:1024	Log all TCP traffic for any IP address and port number to destination IP address (192.168.168.130/24), destination port less than or equal to 1024
Log TCP any any → 192.168.168.130/24 1024:	Log all TCP traffic for any IP address and port number to destination IP address (192.168.168.130/24), destination port greater than or equal to 1024
Log TCP any any → 192.168.168.130/24 ! 1000:1024	Log all TCP traffic for any IP address and the port number to destination IP address (192.168.168.130/24) and destination port number except 1000 to 1024

6.4 Rule Header

(1)   Alert—It will alert for incoming network packets and log in the log files.

(2)   Log—It will log the incoming network packet.

(3)   Pass—This will ignore the network packet.

(4)   Activate—It will alert and then start dynamic rules for inspection.

(5)   Dynamic—It will become idle until activated by an activating rule.

Snort mainly supports three types of protocols—TCP, ICMP, and UDP to detect suspicious behavior. In upcoming releases, it will support other protocols too, such as—IGRP, ARP, RARP, RIP, OSPF. Snort allows both 32-bit IP address and IP address with Classless Inter-domain Routing (CIDR) notation. It also supports NOT (!) operator to exclude the IP address for inspection, example (alert tcp ! 192.168.168.129.0/24 any → 192.168.168.130/24), you can also use multiple IP address using [] (alert tcp ! [192.168.168.129.0/24, 10. 10. 10. 1/24] any → [192.168.168.130/24, 10. 10. 10. 2/24]).

6.5   Rules Options

(1)   Msg—It will display the message for alerts.

(2)   TTL—time to live—It will test the IP header till its TTL value.

(3)   TOS—type of service—It will test the IP header till its TOS value.

(4)   Flags—Snort will test the TCP flags such as—R—RST (reset), A—ACK (acknowledge), P—PSH (push), U—URG (urgent).

(5)   Id—Test the IP header fragment ID field.

(6)   Dsize—For integrity check, it will test the payload size.

(7)   Itype—Test the value of the ICMP type field.

(8)   Session—It is used to extract user data from the TCP session. It has two attributes—printable and all, where "printable" will print only the user would typically trying to print to connect to telnet, FTP, or any other web session (default credentials), and "all" will print non-printable characters with hexadecimal notation.

(9)   Rpc—It will test for any remote procedure call in network traffic.

(10)   Content-list is generally used to block specific traffic with content or pattern that you put in this list.

(11)   Seq—It will test the value for the TCP sequence number field in the packet.

(12)   Ack—It will test the value for TCP acknowledge number field in the packet.

(13)   Logto—This will log the packet to the specified file and not in the standard output file.

(14)   Fragbits—This will inspect any reserved bits in the IP header.

(15)   Content—This is one of the most essential components of Snort, where it inspects specific content in the packet and compares it with the Boyer–Moore pattern matching function. An alert is generated if the match is successful and the content is found anywhere in the packet. Moreover, we can also set the "depth" option for maximum search depth for pattern matching. This could be useful in limiting our search if we know where exactly we should look for this content.

```
alert  tcp  any  any  -> 192.168.168.130/24  111  (content:
"Tor"; depth:20; msg: "Buffer Overflow Detected";)
```

7   Open-Source Host-Based Intrusion Detection System (OSSEC)

As the name specifies, it is a host-based intrusion detection system, which works similarly to Snort, but it will protect your system rather than your network from intruders. It provides a multi-platform, where you can gather information from different operating systems irrespective of their physical structure. With a real-time monitoring system, it can alert the administrator with its configurable and customizable architecture. In addition to that, it checks the integrity of any file or registry- syscheck, also detects harmful malware and provides an efficient way to log the alerts. Furthermore, it has a UNIX rootkits detection script, which can detect any modification to the system state. The architecture of OSSEC has one OSSEC manager, which is a centralized system to collect information from its OSSEC agents (devices that are to be monitored). These agents have log files, commands, and a database of files and registry process status. They are connected with 1514 and 1515 UDP/TCP ports, where 1514 is used for main communication and 1515 is used for the registration process to the manager. All agents could be an individual machines running separately in an isolated environment or could be grouped. OSSEC is available for all major operating systems Windows, Mac, Linux, Solaris, FreeBSD, and virtual machines. Snort analyzes the network packets using sniffing, then it detects and alerts about any possible threat. In contrast to that, OSSEC heavily depends on its log files; from there, it detects and alerts about any possible attack. The architecture of OSSEC with its manager and agent is shown in Fig. 8.

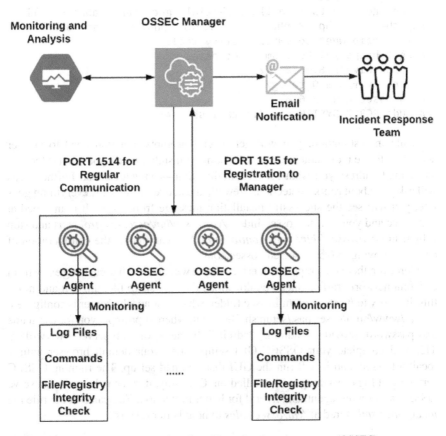

Fig. 8 Architecture of open-source host-based intrusion detection system (OSSEC)

7.1   Installation of OSSEC in Linux

Before installing OSSEC in Ubuntu, we need to update and upgrade our system, to work compatibly with packages. Then, you need an extra library such as *gcc* compiler and other PHP packages to work with OSSEC [15], and then finally, you can download the stable release for OSSEC (https://www.ossec.net/downloads/).

We are working with release 3.6.0—the latest stable release. Once you untar the file, in the ossec folder, there is an install.sh file, you need to run it for a successful installation. There might be some errors pops up, such as fatal error—*"event.h, pcre2.h, zlib.h, opensslv.h—No such file or directory found—error 0 × 5."* You can resolve event.h—by installing libevent-dev, zlib.h can be resolve using—zlib1g-dev, for opensslv.h—you need to install—libz-dev libssl-dev libpcre2-dev libevent-dev build-essential. For pcre2.h—you can download the tar file of pcre2 and paste it to the path—/ossec/src/external.

```
# apt-get update
# apt-get install build-essential gcc make apache2 li-
bapache2-mod-php7.0 php7.0 php7.0-cli php7.0-common apache2-
utils unzip wget sendmail inotify-tools
# wget https:/github.com/ossec/ossec-
hids/archive/3.6.0.tar.gz
# tar -xvzf 3.6.0.tar.gz
# cd ossec-hids-3.6.0
# sudo PCRE2_SYSTEM=no ./install.sh
```

While in installation, you will get several prompts that you need to answer according to the requirements, e.g., what kind of installation do you want ("local"), any e-mail address you want for alert notification—root@localhost. Further steps will ask you about rootkit detection, firewall, and active response. If everything goes fine, you will see the successful installation message from ossec. It is installed at /var/ossec and you can start ossec-hids—*/var/ossec/bin/ossec-control start* and stop it by using */var/ossec/sbin/ossec-control stop*. You can update the configuration of ossec by moving to /var/ossec/etc./ossec.conf.

Moreover, there is a graphical interface where we can visualize each log entry from different network traffic. Once you download the *master.zip* file, unzip it and move this directory to /var/www/html/ossec folder, where our apache server is configured. In /var/www/html/ossec has./setup.sh file, run it, where it prompts you for username and password, provide—"admin," "admin," the name of web server—www-data. This will complete your OSSEC GUI setup. Open your default browser with—localhost/ossec, and it will run the GUI that you had set up. The running OSSEC service will log everything it is installed on. Currently, it is local, but you can serve OSSEC as a server, agent, and hybrid for better protection. You can edit the rule file *local_rules.xml* stored at */var/ossec/rules* to hold better results.

```
# wget https://github.com/ossec/ossec-wui/archive/master.zip
# unzip master.zip
# mv ossec-wui-master /var/www/html/ossec
# cd /var/www/html/ossec
# ./setup.sh
# systemctl restart apache2
```

To test our system, we installed the TOR browser and fired it up, and in few seconds, OSSEC grabbed this traffic by notifying in the browser about the changes that happen in syslog. Figure 9, shows the OSSEC configured web browser with its different notification about the changes occur in the system. If we open the syslog, we can see the user had installed the TOR browser, and this is valuable information from the perspective of network forensics, Fig. 10 shows the syslog.

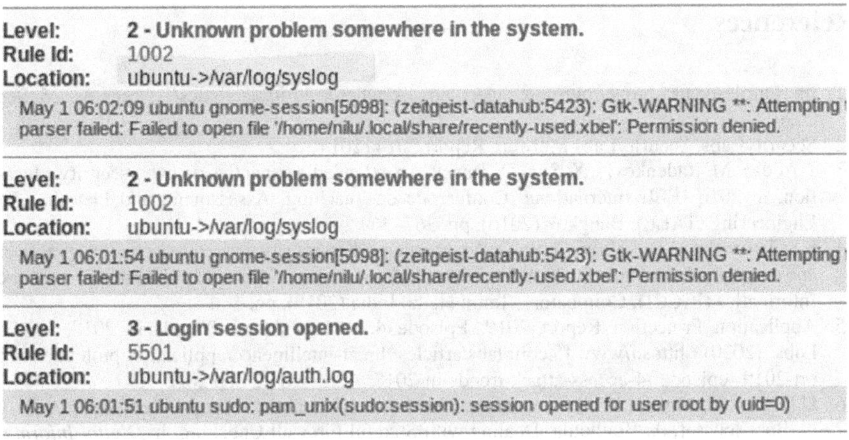

Level:	**2 - Unknown problem somewhere in the system.**
Rule Id:	1002
Location:	ubuntu->/var/log/syslog

May 1 06:02:09 ubuntu gnome-session[5098]: (zeitgeist-datahub:5423): Gtk-WARNING **: Attempting parser failed: Failed to open file '/home/nilu/.local/share/recently-used.xbel': Permission denied.

Level:	**2 - Unknown problem somewhere in the system.**
Rule Id:	1002
Location:	ubuntu->/var/log/syslog

May 1 06:01:54 ubuntu gnome-session[5098]: (zeitgeist-datahub:5423): Gtk-WARNING **: Attempting parser failed: Failed to open file '/home/nilu/.local/share/recently-used.xbel': Permission denied.

Level:	**3 - Login session opened.**
Rule Id:	5501
Location:	ubuntu->/var/log/auth.log

May 1 06:01:51 ubuntu sudo: pam_unix(sudo:session): session opened for user root by (uid=0)

Fig. 9 Graphical user interface of OSSEC Log

```
nautilus-autostart.desktop
May  1 05:57:23 ubuntu org.gnome.zeitgeist.SimpleIndexer[4965]: *
tor-browser.desktop
May  1 05:57:28 ubuntu sm-mta[18148]: 041CvSPl018148: from=<ossec
nsgid=<202005011257.041CvSPl018148@ubuntu>, proto=SMTP, daemon=MT
May  1 05:57:28 ubuntu sm-mta[18152]: 041CvSPl018148: to=<root@lo
nailer=local, pri=30900, dsn=2.0.0, stat=Sent
May  1 05:57:40 ubuntu org.gnome.zeitgeist.SimpleIndexer[4965]: *
nilu/Downloads/tor-browser_en-US/start-tor-browser.desktop
May  1 05:59:09 ubuntu sm-mta[18348]: 041Cx99a018348: from=<ossec
nsgid=<202005011259.041Cx99a018348@ubuntu>, proto=SMTP, daemon=MT
May  1 05:59:10 ubuntu sm-mta[18351]: 041Cx99a018348: to=<root@lo
nailer=local, pri=31267, dsn=2.0.0, stat=Sent
May  1 06:00:53 ubuntu gnome-session[5098]: (zeitgeist-datahub:54
```

Fig. 10 Artifacts of TOR browser in syslog, captured by OSSEC

8 Summary

Though we have a concrete security defense line in terms of firewall, access control, and antivirus, enterprise resources will always be in an active zone of attack. Every day, a vulnerability has been found and gets exploited, the reason is many organizations are not applying effective protective parameters, and hence the intrusion happens. IDS and IPS are very effective in detecting these nuisance activities. By using NIDS and HIDS, we can protect our resources from many severe attacks such as ransomware. The chapter gives a baseline of protective measures that should be applied to any industry or to an individual. Furthermore, it specifies about Snort and OSSEC installation with its rule specification.

References

1. In: Purplesec.us. https://purplesec.us/resources/cyber-security-statistics/. Accessed 26 Sep 2020.
2. SecurityLabs, SecurityLabs Malware Report 2014 (2014)
3. J. Addae, M. Radenkovic, X. Sun, D. Towey, An extended perspective on cybersecurity education, in 2016 IEEE International Conference on Teaching, Assessment, and Learning for Engineering (TALE), Bangkok (2016), pp. 367–369
4. S.S. Tirumala, M.R. Valluri, G. Babu, A survey on cybersecurity awareness concerns, practices and conceptual measures, in 2019 International Conference on Computer Communication and Informatics (ICCCI), Coimbatore, Tamil Nadu, India (2019), pp. 1–6
5. Application Protection Report 2019, Episode 4: Access Attack Trends in 2018. In: F5 Labs (2020). https://www.f5.com/labs/articles/threat-intelligence/application-protection-report-2019--episode--4-access-attack-trends-in-2018. Accessed 22 Sep 2020
6. U.H. Rao, U. Nayak (2014) Firewalls. in The InfoSec Handbook. Apress, Berkeley, CA
7. N. Ben Youssef, A. Bouhoula, Dealing with Stateful Firewall Checking, in *Digital Information and Communication Technology and Its Applications. DICTAP 2011. Communications in Computer and Information Science* ed. by H. Cherifi, J.M. Zain, E. El-Qawasmeh, vol. 166 (Springer, Berlin, 2011)
8. I. Market, Intrusion detection and prevention systems market by solutions & services—2025|markets and markets, in Marketsandmarkets.com. https://www.marketsandmarkets.com/Market-Reports/intrusion-detection-prevention-system-market-199381457.html. Accessed 19 Aug 2020
9. System Intrusion Detection and Prevention, in *Computer Network Security* (Springer, Boston, 2005)
10. C. Panos, C. Xenakis, I. Stavrakakis, An evaluation of anomaly-based intrusion detection engines for mobile Ad hoc networks, In *Trust, Privacy and Security in Digital Business. TrustBus 2011. Lecture Notes in Computer Science* ed. by S. Furnell, C. Lambrinoudakis, G. Pernul, vol. 6863 (Springer, Berlin, 2011)
11. S.R. Snapp, S.E. Smaha, D.M. Teal, T. Grance, The DIDS (distributed intrusion detection system) prototype, in Proceedings of the Summer USENIX Conference (USENIX Association, San Antonio, Texas, 1992), pages 227–233, 8–12 June 1992
12. I. Ahmad, M. Basheri, M.J. Iqbal, A. Rahim, Performance comparison of support vector machine, random forest, and extreme learning machine for intrusion detection, in *IEEE Access*, vol. 6 (2018), pp. 33789–33795
13. Sourcefire (2020), in En.wikipedia.org. https://en.wikipedia.org/wiki/Sourcefire. Accessed 11 Sep 2020

14. K. Salah, A. Kahtani, Performance evaluation comparison of Snort NIDS under Linux and Windows Server. J. Netw. Comput. Appl. **33**(1), 6–15 (2010)
15. J. Vukalović, D. Delija, Advanced persistent threats—detection and defense, in *2015 38th International Convention on Information and Communication Technology, Electronics and Microelectronics (MIPRO)* (Opatija, 2015), pp. 1324–1330

Chapter 7
Introduction to Malware Analysis

1 Introduction of Malware

Nowadays, with increased usage of technology and the Internet, global communications within society are becoming more and more popular. Internet is a free channel, which is vulnerable to different kinds of attacks like data modification attacks, data integrity attacks, etc., under the bigger umbrella of MALicious softWARE, i.e., *MALWARE*. Criminals and intruders of the Deep Web buy malware, modify it, and increase the complexity of code to extend the obfuscation and decrease the possibilities of being detected by antivirus vendors [1]. This prompts different forks or new usage of the same kind of malicious programming that can proliferate out of control. Therefore, the vulnerability of utilizing the Internet is expanded because of the increasing dangers of malware, which get transported inside the system and documents through the Internet.

Malware, in other words, known as vindictive programming, is designed to harm personal computers (PCs), or computer systems, servers, etc. [2]. It is a software program, which denies/disrupts business operations, gathers data for privacy leakage/exploitation, unauthorized access to system resources, and other offensive behavior. Malware is a generic term software professionals use to refer to intrusive, hostile, and annoying software programs. Various kinds of malware families exist, such as viruses, Adware, Spyware, Bots, Trojan, Keylogger, rootkit, phishing, and ransomware. Consistently, one of the organizations, AV-TEST, registers more than 3,50,000 new malignant projects, i.e., malware and potentially unwanted applications (PUA). Then, registered malware is analyzed by its characteristics and ordered based on its spread [3].

Furthermore, representation programs produce the outcomes in charts, which provide insights into current malware. As per the report of Statista, the worldwide detections of newly developed malware applications reached up to 677.66 million in March 2020. Figure 1 shows the growth of malware worldwide from January 2010 to March 2020 [4].

© The Author(s), under exclusive license to Springer Nature Singapore Pte Ltd. 2022
N. Dutta et al., *Cyber Security: Issues and Current Trends*, Studies in Computational Intelligence 995, https://doi.org/10.1007/978-981-16-6597-4_7

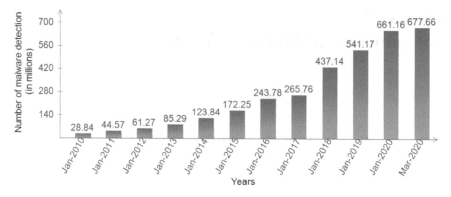

Fig. 1 Growth of malware applications worldwide

In 2020, AV-TEST conducted a test on malware, where a total of 1062.4 million malware appeared. It is expected to increase the malware ratio by approx. 5.65 million by June 2020 [3]. The malware family consists of different varieties of malicious code, such as Trojan and Keylogger. Figure 2 shows the different types of malware, which affect the normal operation of PCs, servers, smartphones, etc. Then, the fight between malware researchers and security analyzers is everlasting due to technological development.

Malware identification comprises a fine model to recognize destructive code. Mostly, malware creators target the system and plan it properly before developing new code and updating the old code with new parts or simply jumbling the system. With a large number of malware cases, limitless malware appears every day, which are almost identical to affect the flow of a system [5]. The following section discusses the varieties of malware in detail.

2   Types of Malware

Figure 2 illustrates the different types of malware, which affect the normal operation of PCs, servers, smartphones, etc. These malware types are as follows [6–17].

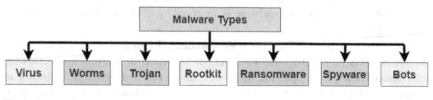

Fig. 2 Types of malware

(i) *Virus*

It is a computer program, which infects another computer program by modifying them to embrace an advanced copy of itself. It is a type of malware, which duplicates itself and spreads to different systems. Many systems can be infected in case viruses attach themselves to various codes and programs. This infection can be used to damage host systems, make botnets, take data, take cash, and render ads [18].

(ii) *Worms*

By exploiting operating vulnerabilities, worms used the network to be spread within the network. It damages the computer system by consuming the bandwidth and overloading the servers. Worms consist of payloads designed to delete and steal data, or worms can create a botnet. Worms and viruses are different. A worm can self-replicate and spread itself, while a virus needs human intervention [18].

(iii) *Trojan horse*

It is commonly known as Trojan, a type of malware that disguises itself as an important document or program to fool clients into downloading/installing it with hidden malicious landscapes. A Trojan can give pernicious remote access to a contaminated PC. Once the system gets infected, attackers can access it and do harmful activities like stealing financial data, data logins, even electronic money, modifying files, installing more malware, and monitoring user activity [18].

(iv) *Rootkit*

This malicious software is designed in such a way that to bypass the security programs while remotely accessing a computer. Once a rootkit is installed in the backend, it will modify the system configuration and attacker access or steal all data. Due to their stealthy operation, it is pretty difficult to prevent, detect, and remove rootkit from an infected system [18]. To detect such types of malware, we need to rely on manual methods, for instance, signature scanning, monitoring system behavior for any irregularity and dump analysis of data storage.

(v) *Spyware*

It is a kind of malware spy on a user's activity without any awareness of the user. It comprises collecting keystrokes, activity monitoring, data harvesting, etc. Spyware frequently has additional capacities like altering the security settings of programs and interfering with the connections between two or multiple programs. Spyware spreads by various factors, for instance, programming vulnerabilities and wrapping itself in Trojans or with real programming [18].

(vi) *Ransomware*

This type of malware restricts the use of computer systems. They encrypt all the data in a computer, popping a message that the client/user needs to pay the malware developer. Otherwise, system access will be recovered or restricted to the limited access to their PC. Ransomware spreads like a usual PC worm, for instance, through a downloaded record or some other delicacy in the system administration [18].

(vii)   **Bots**

It refers to a bug, a defect within the computer software that produces an undesired result. This happens due to human mistakes and mostly exists in a computer program's compilers/source code. The available literature is categorized into (i) minor bugs and (ii) major bugs. The first one influences a program's conduct. On another side of the coin, the major bugs can cause freezing or smashing. Security bugs are one of the most severe kinds of bugs and can permit attackers to evade client verification, take information, etc. On the other hand, bugs can be anticipated with the help of quality control, designer training, and code investigation mechanism [18].

3   Malware Symptoms

Each malware has a different identity of how they are spreading and infecting the computers and networks. All of them have various symptoms to identify whether our systems are infected with malicious code. Some of the symptoms are listed below:

- Expanded CPU utilization.
- Slow Internet browser speeds or PC.
- Issues associated with computer systems.
- Freezing or slamming.
- Delete or modify records.
- The appearance of odd records, work areas, or project symbols.
- Projects killing, running, or reconfiguring themselves.
- Regularly reconfigure or kill firewall and antivirus software.
- Unusual PC behavior.
- Automatically e-mails/messages are being sent without any knowledge of the client.

4   Need of Malware Analysis and Spreading Mechanism

This section highlights the need for malware analysis due to limitless malware cases appearing every day. Then, this section discusses the spreading mechanism, which is prominently used to spread malware and attack/infect the system.

4.1 Need for Malware Analysis

To protect our data from any malicious attack, we need malware analysis. So that, we can prevent and remove malware from our system. Malware analysis has a high effect on the methodology of choosing an ambiguous application. Malware analysis is categorized into two essential arrangements that incorporate dynamic and static techniques [6].

4.2 Malware Spreading Mechanism

This subsection highlights the various mechanisms used to spread malware to attack a system. Figure 3 shows the several spreading mechanisms for malware, which are as follows:

- Boot sector infection: Infect the physical disk of a system by attacking master boot record (MBR).

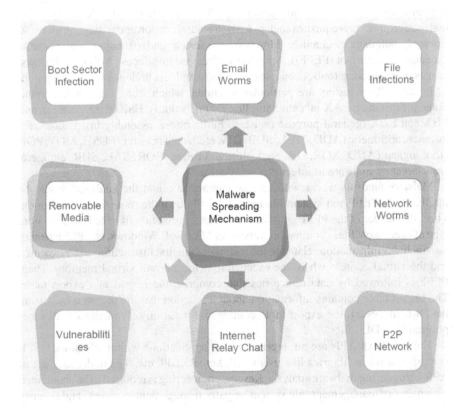

Fig. 3 Malware spreading mechanism

- E-mail worms: Malware attacks the system by sending and receiving worms with e-mails.
- File infections: Parasitic infections and worms, which infect the PC or system via file sharing.
- Network worm: It infects the entire communication channel.
- Peer-to-peer (P2P) network: The PCs or systems get affected due to infected P2P communication networks.
- Internet relay chat.
- Vulnerabilities: Web browser plugins, operating system (OS), adobe reader vulnerabilities are some of the examples, which spread the malware within PCs or systems.
- Removable media: USB drives, pen drives, floppy media, and optical disks also contribute to spreading malware quickly.

5 Malware Analysis Prerequisites

The above discussion till this point is all about what malware is, how it is spreading. Furthermore, the most crucial question is "How to analyze it". To analyze the malware requires prerequisites such as malware classification understanding and ×86 assembly language mechanism. Moreover, file format understanding (e.g., portable executable file format (PEFF)), Application Program Interfaces (APIs) of windows, usages of monitoring tools, debuggers, and disassemblers understanding is required.

In the CPU, registers are particular locations, which store data for manipulations, for instance, EAX (accumulator that returns values), ESP (stack pointer), and EBX and EDX (general-purpose register). Furthermore, assembly instructions, for instance, arithmetic (ADD, MUL, SUB, DIV, etc.), data transfer (PUSH, MOV, POP, etc.), logical (AND, XOR, OR, etc.), shift and rotate (ROR, SAL, SHR, etc.), etc. are a target by malware to infect PCs or systems.

Malware analysis works with reverse engineering, and the analyzer should be familiar with PEFF and system libraries and drill down the root cause of infection. Then, the header of the PEFF contains important information like the linker version, type of executable file, a compatible version of Microsoft Windows, etc. PEFF header also includes information related to the address of the first instructions to be executed and the virtual address where the executable is loaded into virtual memory. Then, PEFF is followed by data directories that comprise the import and export table. The import table contains information about a function that the program calls from the DLL file. Next, the export table contains information on functions (call other programs) in DLL files.

The Windows APIs are an interface to the applications within Windows OS. It comprises a set of libraries like user32.dll, kernel32.dll, etc. The understanding of APIs helps during malware analysis. Reverse engineering is required to be acquainted with memory management, file system, registry management, network, and security APIs.

6 Malware Analysis Environment

While starting with malware analysis, the analyzer should be very careful with malware samples. Furthermore, the malware analysis must be in an environment separated from production and on an isolated network not connected to the public network. Several virtualization software such as VirtualBox and VMWare are feasible solutions to create such an environment for malware analysis.

7 Malware Detection System and Analysis

This section comprises a detailed discussion on malware detection with its analysis. Figure 4 shows the malware detection and analysis system, which consists of malware analysis and malware detection.

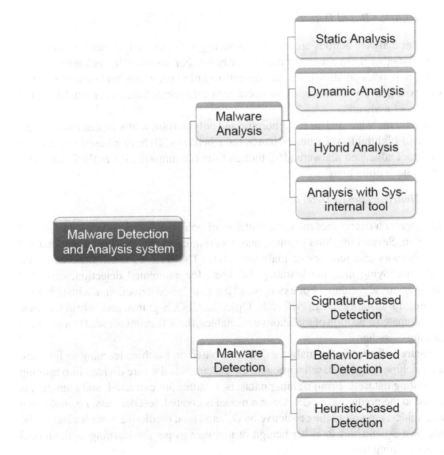

Fig. 4 Malware detection and analysis system

7.1 Malware Detection

It is classified into three categories: (i) signature-based detection, (ii) behavior-based detection, and (iii) heuristic-based detection.

i. Signature-Based Detection

For every program, a signature is unique (a combination of a short byte of strings), and it is used to detect malware in the .exe file, boot file, and memory areas. Some hypotheses state that each version of the malware has the identical signatures.

When malware authors have developed new malware and the system is infected with new malware, the security analyzer stores this new signature in the database. So, whenever new malware appears in a system, we can match the signature with the database. Still, this approach is inefficient as new variations of metamorphic and polymorphic malware will not be detected. Moreover, signature-based malware detection requires a considerable amount of time, people, and money to extract signatures [15].

ii. Behavior-Based Detection

Different malware actions are monitored during malware analysis, and analysts can decide whether these actions consist of malware. For instance, the behavioral operations, network operations, registry operations, file operations, and memory operations exist in every malware. These operations are accomplished via robust dynamic analysis of malware.

Several malware such as metamorphic and polymorphic malware can change their code and highlights yet cannot change their behavior. Behavior-based recognition can detect advanced malware [16], though time consumption for analysis and false alarm rate is quite high.

iii. Heuristic-Based Detection

This approach overcomes the disadvantages of behavior-based and signature-based detection. Several machine learning and data mining techniques are used to extract the behaviors and features of malicious code. The features need to be extracted while employing machine learning classifiers for automated detection, extracted using various algorithms. For example, a Heuristic-based detection mechanism uses features, for instance, DLL, API calls, Op-code, CFG, n-grams, and hybrid features. For automated detection of malicious executables, these features are sent to a machine learning classifier.

Figure 5 illustrates the malware detection based on machine learning techniques. In a machine learning-based approach, the executables data are divided into training and testing datasets. From training datasets, features are extracted, and then data is passed to the predictive model. Once a model is created, test datasets, i.e., unknown executables, are sent to the predictive model, and then predictive models classify the unknown executables to either benign or malware as per the learning of the model during training time.

Training Data

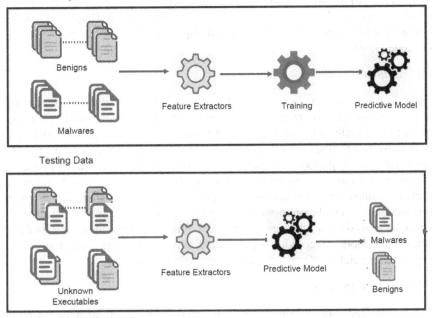

Testing Data

Fig. 5 Malware detection using machine learning

Several machine learning techniques exist to create prediction models such as linear regression, random forest, decision tree, long short-term memory (LSTM), artificial neural network (ANN), support vector machine (SVM), etc. Here, time consumption for malware detection is less, and also, the false positive rate is low [17].

7.2   Malware Analysis

A procedure of deciding the nature, reason, behavior, and usefulness of malware is referred to as *malware analysis*. This procedure is required to identify whether a specific malware is exploiting vulnerabilities or not. As per the existing literature, there are essentially four kinds of malware analysis: (i) static analysis, (ii) dynamic analysis, (iii) hybrid analysis, and (iv) analysis using a sys-internal tool.

(i)   *Static Analysis*

A malware that is analyzed without being executed in a run-time environment is known as static analysis. Here, malicious code needs to be unpacked and decrypted before doing analysis. The Op-code, control flow graph, string signature, and byte-sequence n-grams need to be observed to identify malware. So, various disassembler tools like IDA-Pro and OllyDbg are used for analysis (to observe malicious code), which displays Intel ×86 instructions [7, 8]. Furthermore, OllyDump and LordPE can be used for memory dump and they will display the location of memory through Op-code [9, 10].

(ii) *Dynamic Analysis*

In a safe environment, malware is executed, and the behavior of malware is measured in the dynamic analysis process. To perform dynamic analysis, mostly parameter evaluation, function call monitoring, and information tracking are considered. In a scenario, where malware is accomplished in a virtual environment, it might be conceivable that it performs uniquely like real malware. So it is required to execute the same malware in automated dynamic analysis tools such as TT Analyzer, cuckoo sandbox, and Norman sandbox [11–13]. Report generated via these tools provides details of malicious behavior of the malware and also classified the malware. Dynamic analysis is much more scalable and effective compared to static analysis. Antivirus vendors are receiving huge amounts of malicious code samples every day, requiring artificial intelligence for automatic malware detection.

(iii) *Hybrid Analysis*

It is an amalgamation of both dynamic and static analysis. In the hybrid analysis approach, first static analysis is performed on malicious code then, dynamic analysis is employed for proper detection of malicious behavior of the malware [14–16]. It is comparatively much more effective than static and dynamic analysis as it combines both the analysis approach.

(iv) *Analysis with the Sys-Internal Tool*

The sys-internal tool is generally used for analyzing malware and cleaning the malware. By using this tool, an analyst can understand the impact of malware and malware operations. In addition, this tool provides various malware checking properties like malware signature and DLL files [17–20].

With the increase of advanced attacks techniques, such as spear-phishing attacks, zero days, persistence maintenance, lateral movements, and water holing attacks, malware analysis has high importance as most malware are designed to target cyber-systems. The most evasive techniques like cryptors, packers, obfuscation, etc., are used by modern malware [21–23].

As per the available literature, Advanced Persistent Threat (APT) is a continuous hacking process, which is achieved using automated malware that is left over the targeted system to obtain unauthorized access and stay undetected for a lengthy time period. Table 1 shows a detailed comparison between malware and APT based on various parameters, for instance, target random, signature, evasion, etc. [24, 25].

APT is achieved by specific types of threat actors who are sophisticatedly skilled, highly paid, and well determined to achieve goals. The major goal of APT is to steal data instead of causing damage to the system, for example, Duqu, Flame, and Stuxnet [26–31].

Table 1 Comparative analysis between malware and APT

Particulars	Malware	APT
Target random	Hosts/networks	Specific hosts/networks
Persistent mechanism	Possible	Yes
Signature	Known	Unknown
Covert communication	Possible	Yes
Lateral movement	Possible	Yes
Antivirus detection	High	Low
Firewall/IDS detection	Yes	Very low
Evasion	No	Yes
Threat vector	Generic malware	Zero days

8 Conclusion

Malicious attacks are a prime concern to put PC and correspondence frameworks in danger to harm devices or steal data. Nowadays, malware is a fast-growing threat to the modern computing world, and we do not have a preferred solution to tackle it. Moreover, with the exponential growth of the Internet with various services such as social networks, and cloud storage, malware rapidly multiplies itself, and sometimes it leads to a major attack like botnet attack and others.

This chapter highlights the background of malware, its types, and existing approaches for detection and analysis of the malware. This chapter also discussed the symptoms of malware with the needs of its analysis and spreading mechanism. Here, we highlight the prerequisites and environment for malware analysis. Then, we classified the malware detection and analysis system into different categories as per the available literature. In addition to that, this chapter provides awareness and training in every organization to make the reader understand the fundamentals of malware, its working, and its removal process. Then, the definite order of malicious programming detection and evasion programs is discussed.

In the future, real-life implementation of the malware detection and analysis system will be explored by analyzing a real-time scenario.

References

1. R.S. Pirscoveanu, S.S. Hansen, T.M. Larsen, M. Stevanovic, J.M. Pedersen, A. Czech, Analysis of malware behavior: type classification using machine learning, in *2015 International Conference on Cyber Situational Awareness, Data Analytics and Assessment (CyberSA)* (IEEE, 2015), pp. 1–7

2. L. Liu, B.S. Wang, B. Yu, Q.X. Zhong, Automatic malware classification and new malware detection using machine learning. Front. Inf. Technol. Electron. Eng. **18**(9), 1336–1347 (2017)
3. https://www.av-test.org/en/statistics/malware/. Accessed 20 June 2021
4. https://www.statista.com/statistics/680953/global-malware-volume/. Accessed 20 June 2021
5. S.D. Nikolopoulos, I. Polenakis, A graph-based model for malware detection and classification using system-call groups. Journal of Computer Virology and Hacking Techniques **13**(1), 29–46 (2017)
6. Z. Li, L. Sun, Q. Yan, W. Srisa-an, Z. Chen, Droid classifier: efficient adaptive mining of application-layer header for classifying android malware, in *International Conference on Security and Privacy in Communication Systems* (Springer, Cham, 2016), pp. 597–616
7. IDA Support: Freeware Version, https://www.hexrays.com/products/ida/support/download_freeware.shtml. Accessed 22 April 2021
8. OllyDbg v1.10, http://www.ollydbg.de/. Accessed 22 Apr 2021
9. LordPE—Collaborative RCE Tool Library, http://www.woodmann.com/collaborative/tools/index.php/LordPE. Accessed 22 Apr 2021
10. OllyDump—Collaborative RCE Tool Library, http://www.woodmann.com/collaborative/tools/index.php/OllyDump. Accessed 22 Apr 2021
11. cuckoosandbox-Automated Malware Analysis, cuckoosandbox.org, https://www.cuckoosandbox.org. Accessed 24 Apr 2021
12. Norman|Antivirus & Security Software for Home & Business, https://www.norman.com/en-ww/homepage. Accessed 24 Apr 2021
13. TTAnalyzer,   Nsftele.com,   http://www.nsftele.com/NSF%20nostalgy/TTAnalyzer.htm. Accessed 24 Apr 2021
14. D. Uppal, V. Mehra, V. Verma, Basic survey on malware analysis, tools and techniques. Int. J. Comput. Sci. Appl. (IJCSA) **4**(1), 103 (2014)
15. J. Bergeron, M. Debbabi, J. Desharnais, M.M. Erhioui, Y. Lavoie, N. Tawbi, Static detection of malicious code in executable programs. Int. J. Req. Eng. **2001**(184–189), 79 (2001)
16. W. Liu, P. Ren, K. Liu, H.X. Duan, Behavior-based malware analysis and detection, in *2011 First International Workshop on Complexity and Data Mining* (IEEE, 2011), pp. 39–42
17. Z. Bazrafshan, H. Hashemi, S.M.H. Fard, A. Hamzeh, A survey on heuristic malware detection techniques, in *The 5th Conference on Information and Knowledge Technology* (IEEE, 2013), pp. 113–120
18. N. Dutta, K. Tanchak, K. Delvadia, Modern methods for analyzing malware targeting control systems, in *Recent Developments on Industrial Control Systems Resilience* (Springer, Cham, 2020), pp. 135–150
19. A. Ray, A. Nath, Introduction to malware and malware analysis: a brief overview. Int. J. **4**(10) (2016)
20. A. Kumari, S. Tanwar, Secure data analytics for smart grid systems in a sustainable smart city: challenges, solutions, and future directions. Sustain. Comput. Inform. Syst. **28**, 100427 (2020)
21. A. Kumari, R. Gupta, S. Tanwar, Amalgamation of blockchain and IoT for smart cities underlying 6G communication: a comprehensive review. Comput. Commun. (2021)
22. M. Wazid, A.K. Das, J.J. Rodrigues, S. Shetty, Y. Park, IoMT malware detection approaches: analysis and research challenges. IEEE Access **7**, 182459–182476 (2019)
23. Y. Pan, X. Ge, C. Fang, Y. Fan, A systematic literature review of android malware detection using static analysis. IEEE Access **8**, 116363–116379 (2020)
24. A.D. Schmidt, R. Bye, H.G. Schmidt, J. Clausen, O. Kiraz, K.A. Yuksel, S.A. Camtepe, S. Albayrak, Static analysis of executables for collaborative malware detection on Android, in *2009 IEEE International Conference on Communications* (IEEE, 2009), pp. 1–5
25. X. Luo, J. Li, W. Wang, Y. Gao, W. Zhao, Towards improving detection performance for malware with correntropy-based deep learning method. Digital Commun. Netw. (2021)
26. S. Euh, H. Lee, D. Kim, D. Hwang, Comparative analysis of low-dimensional features and tree-based ensembles for malware detection systems. IEEE Access **8**, 76796–76808 (2020)
27. H. Naeem, F. Ullah, M.R. Naeem, S. Khalid, D. Vasan, S. Jabbar, S. Saeed, Malware detection in industrial internet of things based on hybrid image visualization and deep learning model. Ad Hoc Netw. **105**, 102154 (2020)

28. M.K. Alzaylaee, S.Y. Yerima, S. Sezer, DL-droid: deep learning based android malware detection using real devices. Comput. Secur. **89**, 101663 (2020)
29. A. Kumari, S. Tanwar, A secure data analytics scheme for multimedia communication in a decentralized smart grid. Multimed. Tools Appl. 1–26 (2021)
30. S.S. Chakkaravarthy, D. Sangeetha, V. Vaidehi, A Survey on malware analysis and mitigation techniques. Comput. Sci. Rev. **32**, 1–23 (2019)
31. M. Wagner, A. Rind, N. Thür, W. Aigner, A knowledge-assisted visual malware analysis system: design, validation, and reflection of KAMAS. Comput. Secur. **67**, 1–15 (2017)

Chapter 8
Design of a Virtual Cybersecurity Lab

1 Introduction of Cybersecurity

With the fast-growing economy, the cyberworld in recent years has been changed drastically, continuously rising and expanding; one of the main reasons for this drastic change is the fast Internet connection (signifies access to anyone at any time). The cyberworld is a critical area where known and unknown things happen within no time. After a successful cyberattack occurs in an organization, there is very little time to understand what happened and what we could do.

There are 4.13 billion worldwide Internet users till 2019. Among them, China has the highest number of Internet users (854 million users). These numbers themselves say that anyone can access and everything around the globe. This property of Internet usage can be beneficial for a third party (an attacker) who wants to access the user's system. An attacker will manage to get into the system once he gathers information about a user, user's account, and Internet usage patterns. This way, an attacker can access the assets which belong to a user that now belongs to him. Some examples of advanced attacks are GhostNet, Moonlight Maze, Shadow Network (attack on India by China), cyberwarfare, etc. Unfortunately, there is no such restricted boundary to stop these attacks. Instead, it can be prevented by using a preventive measure that is updated every second.

There is a constant fear while accessing anything over the Internet, even a single website whose digital certificate is not authenticated or expired, people get frightened in opening such sites, though they are not harmful, until and unless people download something from such sources. There are cases around the world where people have a phobia of the "Internet", where anxiety and fear play an essential role in causing significant disability in a person's life. A person's life is heavily dependent on the Internet. Hence, the fear of getting troubled by cybercriminals, privacy issues, fear of losing their personal belongings, computer viruses can corrupt files, and attackers can trick us into getting sensitive information are some of the barriers where we freely cannot access the Internet. Those questions keep haunting us while surfing the Internet [1].

N. Dutta et al., *Cyber Security: Issues and Current Trends*, Studies in Computational Intelligence 995, https://doi.org/10.1007/978-981-16-6597-4_8

Another way we can stop such wicked acts is via training. We should know our system properly, develop an understanding of the latest defenses and threats, and apply patches and updates whenever it launches [2]. Superpower nations such as the USA are only spending a billion dollars per year to protect their digital assets from third-party intruders, and these dollars are continuously rising per year to secure digital systems. These huge numbers are somehow wasted in cybersecurity, which shows that cybersecurity is the top priority for nation growth, explained by Mendonsa in an interview at Information Security Media Group [3, 4].

There is a need for awareness of cybersecurity among people, where everyone can understand its terminologies, abstract systems, flaws, and how to overcome these flaws. Awareness is about getting knowledgeable and getting combined with the attitude and practice that can protect our information assets. Being aware also signifies understanding of how those attacks happen, from which vulnerability or loophole leads to an attack, and the steps taken to stop those attacks. In a big organization, awareness programs, various webinars, and seminars are conducted so that people could understand cybersecurity [5, 6]. Machine learning algorithms are the promising solution in order to investigate an online threats or vulnerabilities. Several research works have been done in this regard, such as Sedjelmaci et al. [1] proposed Artificial Intelligence (AI)-based mechanism for cyber-physical systems. Then, Babiceanu et al. [2] presented a detailed analysis of cyberphysical systems based on virtualization and big data. In [3], virtual remote laboratories are emulated and evaluated for cybersecurity. Moreover, various AI-based mechanisms have been proposed for cybersecurity [4–9].

At IBM, employees get educated and trained on various platforms to face an anonymous attack and defend it. This way a training could be beneficial to handle an attack in the near future. There is a shortage of cybersecurity specialists in the IT field, and degree courses offered in universities are getting scarce. Hence, we need to encourage the IT field to get trained by which they can protect the national infrastructure from a severe attack [7–9].

2 Tools for Cybersecurity

This chapter comprises various techniques and mechanisms, which lead to setting up a cyberlab using various tools to stop any mischievous act. These tools are as follows [1, 10, 11]:

(i) *Virtual Machines*

Physical machines are prone to several attacks via malware and network intrusions. It is easy to attack such machines as they have many attack vectors in the form of software and hardware. Certainly, less effort is required for an attacker to attack the CPU or any software residing in the operating system (OS) compared to the virtual machine.

"Virtuality" is defined as something that possesses an attribute without sharing its physical form. This definition encourages the IT field to create virtual machines, which are somehow challenging for any cybercriminal to attack. Virtual machines produce a virtual environment with a virtual computer with their CPU, network interface, memory, and storage over any physical hardware system. It takes a physical computer's resources and gets converted into virtual ones using specialized software, hiding the underlying procedure (an abstract mechanism). One beautiful example of a virtual machine is "cloud computing", where we can store our resources in the virtual cloud instead of physical storage, and it replaces the need for an external or physical hard drive. Virtualization leads to two critical concepts—(a) Simulation and (b) Emulation.

(a) **Simulation**

A system can be simulated using simulation software, where the system's internal logic and components will also be simulated. The simulation gives the flexibility to run programs that are not possible to run directly on that specific system but can run using simulation. *Mac—Xcode* and *MATLAB–Simulink* are the best examples of simulations to test program code design.

Mobile applications should be run on mobile. However, it must be tested multiple times before handing it over to the client. It is a very tidy process if we take the application from the desktop and put it into mobile multiple times for testing. Hence, *Mac-Xcode* is used to test the design of any mobile application on desktop computers. Similarly, *MATLAB–Simulink* provides a simulated environment to model, design, and analyze multi-domain dynamic systems.

(b) **Emulation**

In contrast to simulation, emulation runs the hardware or software components on a system, but not their internal logic. *Android Emulator* is the best example of emulation to test and design code for any Android application.

3 Virtualization for Cybersecurity

Virtualization is a bridge between hardware and software components. When some software gets incompatibility issues and cannot run on that specific system, virtualization resolves this issue by creating a compatibility layer to run that incompatible program on the unsuited system. Moreover, a concrete example of virtualization is thatLinux commands cannot be run in Windows OS, but certainly a platform has been developed, i.e., a "Cygwin" platform to run those commands. Similarly, window programs (.exe) cannot run in a Linux environment, so an emergent platform "Wine" is used in Linux to run window programs. Each virtualization component is discussed in detail as follows [11, 12].

(i)   *Hardware Virtualization*

It refers to the technology that makes it possible for any hardware component to run on a system using virtual software regardless of their physical forms, such as virtual machines, VMware, and VirtualBox. Here, physical machines are called the host, and virtual machines are referred to as guest systems. As much as a guest OS can run on one host machine, this can be done using one abstraction layer known as hypervisor, also called "Virtual Machine Monitor". There are two types of hypervisors—(i) Type 1 (native or bare-metal hypervisors) and (ii) Type 2 (hosted hypervisors). Figure 1 shows the workflow of both the hypervisors type and a detailed discussion on each type of hypervisor H is as follows.

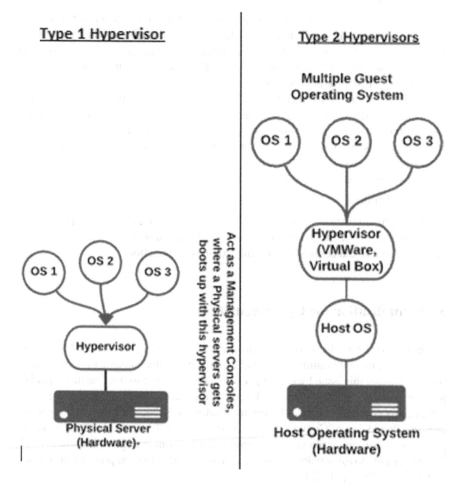

Fig. 1 Workflow of hypervisors

- *Type 1 hypervisors*—These hypervisors run directly on host hardware to control and manage the guest OS using an external console (located off-premises). Microsoft Hyper-V, XCP-ng, VMWare ESXi, and Power Hypervisors are some of the modern hypervisors that are currently available in the market [10].
- *Type 2 hypervisors*—It provides flexibility to run on any conventional OS. One host OS can run multiple guest OS. VMWare Workstation, VMWare Player, Quick Emulator (QEMU), and VirtualBox are some of the examples of type 2 hypervisors.

(ii) *Software Virtualization*

It is also known as application virtualization, which can run software from a remote server situated at any geographical location. The application gets isolated to protect from malware and removed completely from the system. This application can be centrally managed and controlled.

(iii) *Storage Virtualization*

It provides a storage pool by collecting all organization's storage resources in one place, for instance, hard drives, flash, and tape drives. Storage virtualization can be done using software and abstraction layers, which combine all storage resources centrally. Some of the storage virtualization vendors are Datacore SANsymphony-V10, Dell EMC Fast X, and Dell EMC VPLEX.

(iv) *Network Virtualization*

It provides a storage area network, which gives a high-speed network using block-level network access to storage. It gathers storage resources from a heterogeneous system and combines them into a virtual storage pool, for example, VMWare NSX, Nuage Networks Virtualized Services Platform (VSP), Big Switch Network Big Cloud Fabric, and Anuta Networks NCX.

There are various other virtualization mechanisms like server virtualization and desktop virtualization to handle cyberattacks. However, hardware virtualization is the top prior virtualization for cybersecurity, among others, as an environment is required to test malware, security breaches, network attacks, and other intrusions, which is impossible in a physical machine. Therefore, a virtual system could be useful to restore to the point where things were stable, i.e., restore point before a malware attack. The two most popular virtual machines are VMWare and VirtualBox, widely adopted by various organizations and individuals for training purposes.

Figure 2 shows the recent statistics of Spiceworks, which highlight the current state of virtualization from top vendors like VMware, HPE StoreVirtual VSA, etc. [11]. Any of the virtual machines can be used for training purposes, but for demonstration purposes, this chapter considers VMware, and it consists of two versions: (i) Workstation (Premium) and (ii) Player (Free).

The workstation is better than the player as it has some additional features, such as snapshots, cloning, and remote connections to the cloud. In addition, anything can be executed on a virtual machine that is reflected in our physical system.

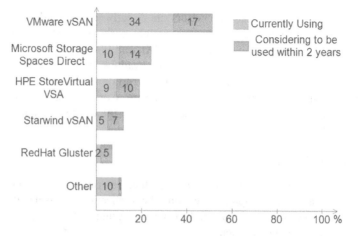

Fig. 2 Different vendors for virtualizations (in %) [11]

4 Installation and Configuration of VMWare Workstation

For VMWare Workstation, the physical machine should be configured with 4 GB
RAM, 2 GHz Processor with virtual technology-enabled from a boot, 200 MB, and
more disk space for a workstation to install an application. Following are the steps
to configure the environment in VMware.

- A fresh copy of the workstation can be downloaded and installed from [13].
- Next, a fresh.iso copy of any Debian system (preferably Kali) and Windows is
  required to make an attacker and a target system.
- In Workstation, load both OS one by one, for 4 GB RAM atleast 1.5 or 2 GB is
  required for a virtual machine, otherwise for 6 GB RAM, need to use at least 2 GB
  for the virtual machine. If the host OS has four processors, then it requires 2 and
  2 cores per processor. (It will vary according to the needs and the capability of
  the host machine). At least 40 GB of disk space is allocated, as several tools and
  libraries will be installed to understand the underlying structure of cybersecurity.
- The network connection should be in network address translation (NAT) format,
  and the host Internet connection is used to connect to the Internet. In the case of
  NAT, an Ethernet connection is visible in the guest machine even Wi-Fi is used
  in a host machine. To get a Wi-Fi connection in a virtual machine, an external
  adapter must be purchased, for example, Edimax Wireless 802 0.11 bgn Nano
  USB Adapter and TP-Link N150 USB Wi-Fi Adapter with SoftAP Mode. A
  virtual machine has drivers for host physical devices such as CPU, RAM, and
  camera, but it does not have any driver for a network device; therefore, an external
  adapter is needed to have a Wi-Fi connection.

Then, start up the virtual machine for installation of specific OS with simple steps.
(Remember to say "Yes" at GRUB installation).

5 Network Modes in Virtual Machines

- *Network Address Translation (NAT)*—It creates a virtual router in a host, which is a subnet on the virtual machine. Then, all traffic to/from the subnet is going to be forwarded by the host. Thus, it gives flexibility for all virtual machines to communicate with each other and access the Internet. (IP address on all VM is different from a host).
- *Host-Only Network*—It enables the network interface controller (NIC) to communicate with the host and other virtual machines. The disadvantage is that it cannot access the host's physical NIC and hence cannot access the Internet, similar to a loopback interface.
- *Bridged Network*—It is more complex, where virtual machines have one IP address on the same domain of host, and it will behave as if it is another machine on the same network where a host is residing. It will set up a device driver in the host machine to filter data from the host network adapter; this filter is known as a "net filter". This allows a virtual machine to intercept data from the physical machine with injecting functionality. From the host perspective, it looks like a guest is connected to the same network with network cables (though it is not). Table 1 shows the details of important networking modes [14].

6 Cybersecurity and Various Attacks

This section highlights the analysis of various attacks to obtain cybersecurity. Cybersecurity requires analysis for various attacks, which are targeting the system. Some of the attacks are as follows [15].

- Information gathering
- Password attacks
- Wireless attacks
- Exploitation tools
- Sniffing and spoofing
- Web hacking

Table 1 Outline of networking modes

Mode	VM ← Host	VM → Host	VM1 ↔ VM2	VM ← Net/LAN	VM → Net/LAN
Host-only	√	√	√	–	–
Internal	–	–	√	–	–
Bridged	√	√	√	√	√
NAT	Port forward	√	–	Port forward	√
NAT service	Port forward	√	√	Port forward	√

- Post exploitation
- Recon
- Reporting.

The next section discusses the defense mechanism to handle various types of attacks (discussed above).

7   Defense Strategies Against Various Attacks

In the scenario, when an attack happens, very little time organizations get to stop that attack, no matter how strong and effective defense strategies the organization has applied. Intruders will surely manage to compromise it, and no such strategy exists, which is completely resistant to an attack. Nevertheless, there are still some ways to help systems be protected and recover from an attack. Some of the examples are as follows.

(i)   **Malware Attack**

Malware analysis can be done in three ways: static, dynamic, and low-level analysis. The static analysis deals with malicious source code and analyzes the code properties to understand the malware behavior. In contrast, dynamic analysis has varieties of tools to understand the actual behavior of the malware, where binary can be executed in a sandbox environment. Then, low-level analysis is the one to provide information when something is missed out by static and dynamic analysis. To understand it, let us consider an unknown malware that has attacked a system, it gets recognized, and then, an automated tool such as process hacker or CaptureBat is required to log every single event that binary has.

Working of CaptureBat

- CaptureBat is a behavioral analysis tool for Win32 OS. It monitors the state of a system when an executable or application gets executed, by which an analyst can see the insights into how an application gets operated.
- In the Windows virtual machine—create a "temp" directory—C:\temp. Then, put any infected file (can download from internet) such as infectedtest.txt inside the temp directory. [15].
- Run command prompt as "Run as administrator".

     # CaptureBAT.exe –c –n –l C:\temp\infectedtest.txt.

- Run the malware, and check the infectedtest.txt file to analyze the binary. Figure 3 shows the outcome of the result after analyzing the infectedtest.txt file [16–18].

In the infectedtest.txt file, there are a lot of entries that this binary had created, and certain entries created and deleted in the registry. Figure 4 shows all the registry entries with timestamps.

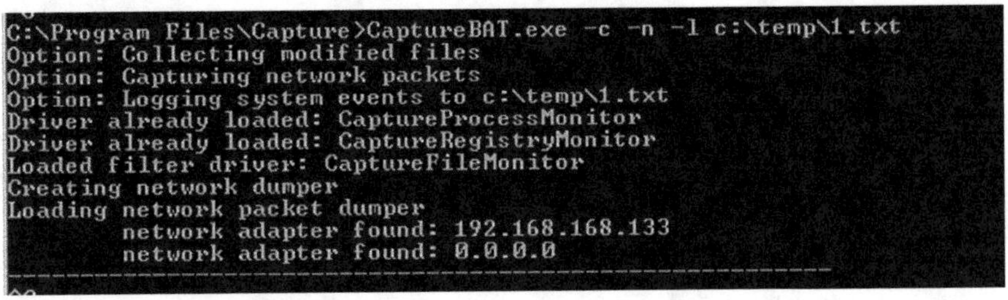

```
C:\Program Files\Capture>CaptureBAT.exe -c -n -l c:\temp\1.txt
Option: Collecting modified files
Option: Capturing network packets
Option: Logging system events to c:\temp\1.txt
Driver already loaded: CaptureProcessMonitor
Driver already loaded: CaptureRegistryMonitor
Loaded filter driver: CaptureFileMonitor
Creating network dumper
Loading network packet dumper
        network adapter found: 192.168.168.133
        network adapter found: 0.0.0.0
```

Fig. 3 CaptureBat execution on cryptovirus

8:45:39.310","registry","SetValueKey","C:\Users\NileshPC\Downloads\CryptoLocke r_22Jan2014\1002.exe","HKCU\Software\935E8D8E0E\Keys"

8:45:39.310","registry","SetValueKey","C:\Users\NileshPC\Downloads\CryptoLocke r_22Jan2014\1002.exe","HKCU\Software\935E8D8E0E\Files"

8:45:39.310","registry","SetValueKey","C:\Users\NileshPC\Downloads\CryptoLocke r_22Jan2014\1002.exe","HKCU\Software\Microsoft\Windows\CurrentVersion\Run\9 35E8D8E0E"

8:45:39.310","registry","SetValueKey","C:\Users\NileshPC\Downloads\CryptoLocke r_22Jan2014\1002.exe","HKCU\Software\Microsoft\Windows\CurrentVersion\RunO nce*935E8D8E0E"

8:45:49.322","file","Write","C:\Users\NileshPC\Downloads\CryptoLocker_22Jan201 4\1002.exe","C:\Users\NileshPC\AppData\Roaming\935E8D8E0E.exe"

8:45:42.471","registry","SetValueKey","C:\Users\NileshPC\Downloads\CryptoLocke r_22Jan2014\1003.exe","HKCU\Software\Microsoft\Windows\CurrentVersion\Run\9 35E8D8E0E"

8:45:42.471","registry","SetValueKey","C:\Users\NileshPC\Downloads\CryptoLocke r_22Jan2014\1003.exe","HKCU\Software\Microsoft\Windows\CurrentVersion\RunO nce*935E8D8E0E"

8:45:49.510","process","created","C:\Users\NileshPC\Downloads\CryptoLocker_22Ja n2014\1002.exe","C:\Users\NileshPC\AppData\Roaming\935E8D8E0E.exe"

Fig. 4 Registry entries for cryptovirus

These are some of the log entries of cryptobinary, which are deleting some of the registries and simultaneously creating and writing it with the value— 935E8D8E0E. Hence, it has created 935E8D8E0E.exe, which is a cryptovirus detected by VirusTotal.

Figure 5 shows the hexadecimal of the sample cryptovirus, and Fig. 6 demonstrates that it is written in.NET language with version 2.0 framework. Then, it is BSJB

61
/73

Community
Score

(!) 61 engines detected this file

5291232b297dfcb56f88b020ec7b896728f139b98cef7ab33d4f84c85a06d553
Microsoft Windows Auto Update.exe

assembly   detect-debug-environment   direct-cpu-clock-access   peexe   r

| DETECTION | DETAILS | RELATIONS | BEHAVIOR | COMMUNITY 10+ |

Ad-Aware   (!) Gen:Variant.Ransom.Blocker.3

AhnLab-V3   (!) Trojan/Win32.Blocker.R185819

Fig. 5 Hexadecimal of cryptovirus

```
 ▯ Ç_Tˢu  .a.¿åoõ  W▯ ý.ó  ÿO▯ Ñ.}°÷W=yɜW<~å
èöûá÷Ôþ.</ᴎvîùýOᵃúiY~  .Ë_  3äy>   ▮▯÷▯ ÿî
¿Fú{FÞùÑósöᴎüSY^W{7¶uÑóÿÅçÿ.BSJB......
......v2.0.50727......t....f..#~..tf..
8s..#Strings....¬Ù..ä...#US. Ú......#G
UID... Ú̱..ì...#Blob................
WS¢......ú.3........ᴎ...Ù...ᵃ...ᴎ...û.
..a...........8...................
. . . . . . . . . . . . . . . . . . . . . . . . . . . . . . . . . . .
        - -      -  -        -      - -     - -
```

Fig. 6 Hexadecimal of cryptovirus

(marked in yellow color in Fig. 6), and when converted to hex, it is 42 53 4a 42, which is the first four letters of the names of four persons who worked on the metadata engine.

Some of the extensions that this crypto has used also used in the digital ledger technology to secure transactions [19, 20].

Table 2 Wi-Fi security rank

Wi-Fi encryption standard	Rank
WPA2 + AES	1 (best)
WPA + AES	2
WPA + TKIP/AES	3 (at security risk)
WEP	4
Open	5 (bad)

(ii) Network Intrusion

There is a constant fear of intrusion attacks in any enterprise. Wired connection is secured by the enterprise premises using security checks like cameras, firewalls, and secured network premises. Wireless connection (Wi-Fi) has five layers of the encryption standard, and many organizations' Wi-Fi falls in Rank 2 and 3, which are easy to crack. Table 2 shows the various Wi-Fi encryption standards and their rank for cybersecurity [21].

With the help of a proper intrusion detection system, network and network services can be secure, for example, Snort, an open-source network intrusion detection system, and OSSEC, a host-based intrusion detection system. Furthermore, there are other ways to stop attacks by creating specific rules for the attack.

- Rule for Malware Detection

alert tcp any any → any any (msg: "Possible Trojan Detected—IRCBot"; flow: to server, established; content: "AAC6F603"; ttl:128; sid: 1000255; rev: 7);

- Rule for Client Accessing Tor Network

alert tcp $HOME_NET any → $EXTERNAL_NET 80,443,9001,9030 (msg: "Tor Network Detected"; classtype: policy-violation; sid: 50009;

- The rule for DNS Zone Transfer

Signature alert tcp $EXTERNAL_NET any → $HOME_NET 53 (msg: "DNS zone transfer"; flow: to_server, established; offset: 15; reference: cve, 1999–0532; classtype: attempted-recon; sid: 255; rev: 11).

- **Website Attacks**

Most of the attacks are application-based, around 70% of the total attacks (mostly website services). Open Web Application Security Project (OWASP) keeps a keen eye on application attacks, and every year they publish top ten web attacks [16, 17]. Some of the most well-known attacks like SQL injection, authentication failure, and cross-site scripting occur every year, and organizations lose millions of dollars because of these attacks. To prevent the attacks mentioned above, below listed actions need to be done, which is as follows:

- Encrypt database tables and restrict access to database servers. Use always parameterized SQL queries to stop SQL injection attacks.
- To prevent cross-site scripting, always encode HTML before it is inserted into the database.
- Use proper protocols while deploying services, never use HTTP, instead use HTTPS.
- Validation is the key to remove errors from the website, use proper validation at each input, and test every entry multiple times with different combinations.
- Logs are crucial at the application layer, keeping a record of everything that happened with the application to track any ambiguity, detect, and monitor any attack.
- All applications should be designed and executed as non-root users and should not give administrator privileges to any normal user. The application should run normally, not with "Run as Administrator" in Windows and Root in Linux.
- Fix any broken authentication by changing the default passwords, use a strategic password, secure cookies, and sessions, limit login attempts, etc. [18, 22].

8 Case Study on Website Attacks

Considering a HTTP website, for example - http://ignou.ac.in/ website which is not secure, as it is HTTP only. Next, we will analyze the website traffic and its content in Wireshark Sniffer [23, 24]. Here, we will start sniffer and parallelly submit a form in the ignou website. The values that we have filled out, can be seen in sniffer output. From the security perspective this is outrageous because if an intruder with this sniffer resides between client and server, the intruder can get all input values. Figure 7 shows, a student registration form that we have filled out, and a Wireshark (Sniffer) that captures all the input values (marked in green).

There are various other techniques to prevent attacks, and one well-known technique is "honeypots". It is a mechanism to detect and capture the unauthorized access of the attacker, where a resource file would be made that looks legitimate. Still, it is an isolated and monitored system to trace and capture the attacker. Thus, it is pretty similar to an intrusion detection system with different aims.

The most significant disadvantage of any intrusion detection system is that it gives false positive alarms, whereas in honeypots there is a less possible chance of false alarms [25, 26].

Honeypot has various capabilities such as IP grabber, DNS and host gathering, geolocation and MAC addresses, TCP port scanner, and an efficient fuzzer. For example, Fig. 8 shows that the honeypot is at IP: 192.168.168.132. Once an attacker accesses this IP with port 80, the honeypot grabs that intrusion with its IP addresses (highlighted in yellow color) then notifies the system administrator. This intrusion detection is done in the Kali virtual machine, which installed Pentbox along with a honeypot in it [20, 27].

ENROLLMENT
NUMBER: 123456789

PROGRAMME: ACPDM   ⌄

DATE OF BIRTH: 06/04/2002   🗓

                 Login

```
                    HTTP      210 HTTP/1.1 200 OK   (PNG)
14.139.40.18        HTTP      814 POST /changeadmdata/AdmissionStatusNew.ASP HTTP/1.1
                    HTTP      1227 HTTP/1.1 200 OK  (text/html)
61 3^ ^1 6^ ^1 65 34 37    3 65 64 64 61 ^^ 35 64    a81c1e47 3edda85d
^^ ^^ ^' 5a  . /5 )a /'  /5 5^ 6/ n' ^'  ^' 74       96=j2umt s0gokebt
6f ^  .  61 . ^P  3^ .. ^  ^^ ^  . ( 1                orjra0p5 8hcp3···
0^ :  ,2 4^    ^^   ^    ^ 38 : )                     ·EnrNo=1 23456789
^^  2 ^'        ^ 3^   '           ( )                &program =ACPDM&d
^   ^   ^    46 ^   3^ ^  ^ : )                       ob=06%2F 04%2F200
.  . 75 ^^ ^  ) 74   ^ ^f  69                         2&Submit =Login
```

Fig. 7 Student registration form with Wireshark Sniffer

```
 INTRUSION ATTEMPT DETECTED! from 192.168.168.133:49174 (2020-06-21 07:19:
11 -0400)
------------------------------
GET /favicon.ico HTTP/1.1
Host: 192.168.168.132
Connection: keep-alive
User-Agent: Mozilla/5.0 (Windows NT 6.1; ) AppleWebKit/537.36 (KHTML, like
Gecko) Chrome/83.0.4103.106 Safari/537.36
Accept: image/webp,image/apng,image/*,*/*;q=0.8
Referer: http://192.168.168.132/
Accept-Encoding: gzip, deflate
Accept-Language: en-GB,en-US;q=0.9,en;q=0.8
```

Fig. 8 Intrusion detection using honeypot

There are many other defense strategies available for system and PC protection, such as training, awareness, and updating the cyberknowledge of everyone (who is using the system) to stop cybercrime. Although it does not protect from the root, indeed, it is a big step to protect nations' digital infrastructure and saves millions of resources for the sustainable development of smart cities [28–32].

9 Conclusion

Over the years, several cyberattacks have damaged the cyberphysical system, where consequences and severity of attacks vary. Without taking prevention for cybersecurity, it will be more dangerous in the future as invaders get more experienced, sophisticated, and malicious. This chapter presented the background of cybersecurity, identification, and analysis of cybersecurity methods. Then, virtual machines, network mode of virtual machines, and defense mechanisms against various attacks are discussed in-depth.

This chapter described the steps to create a cyberlab with varied attributes that could provide a strict boundary against attacks. Further, several software tools and libraries are discussed, which would be helpful in the creating lab. On top of that, a virtual machine is the main focal point with a case study on website attacks. This case study presented intrusion detection using honeypot tools.

References

1. H. Sedjelmaci, F. Guenab, S.M. Senouci, H. Moustafa, J. Liu, S. Han, Cybersecurity based on artificial intelligence for cyber-physical systems. IEEE Netw. **34**(3), 6–7 (2020)
2. R.F. Babiceanu, R. Seker, Big Data and virtualization for manufacturing cyber-physical systems: a survey of the current status and future outlook. Comput. Ind. **81**, 128–137 (2016)
3. A. Robles-Gómez, L. Tobarra, R. Pastor-Vargas, R. Hernández, J. Cano, Emulating and evaluating virtual remote laboratories for cybersecurity. Sensors **20**(11), 3011 (2020)
4. A.B. Nassif, M.A. Talib, Q. Nassir, H. Albadani, F.D. Albab, Machine learning for cloud security: a systematic review. IEEE Access (2021)
5. www.ncbi.nlm.nih.gov/pmc/articles/PMC6241174/. Accessed 20 Jun 2021
6. T. Le, A survey of live virtual machine migration techniques. Comput. Sci. Rev. **38**, 100304 (2020)
7. https://niccs.cisa.gov/about-niccs/national-cybersecurity-awareness-month-2019.   Accessed 25 Jun 2021
8. D. Mishra, P. Kulkarni, A survey of memory management techniques in virtualized systems. Comput. Sci. Rev. **29**, 56–73 (2018)
9. N. Norouzi, G. Bruder, B. Belna, S. Mutter, D. Turgut, G. Welch, A systematic review of the convergence of augmented reality, intelligent virtual agents, and the internet of things. Artif. Intell. IoT, 1–24 (2019)
10. https://en.wikipedia.org/wiki/Comparison_of_platform_virtualization_software. Accessed 21 Jun 2021
11. M. Banerjee, J. Lee, K.K.R. Choo, A blockchain future for internet of things security: a position paper. Digit. Commun. Netw. **4**(3), 149–160 (2018)
12. https://www.spiceworks.com/marketing/reports/state-of-virtualization/. Accessed 21 Jun 2021
13. www.vmware.com/in/products/workstation-pro.html. Accessed 18 May 2021
14. https://www.virtualbox.org/manual/ch06.html. Accessed 18 Jun 2021
15. https://github.com/ytisf/theZoo/blob/master/malwares/Binaries/CryptoLocker_22Jan2014/CryptoLocker_22Jan2014.zip. Accessed 10 Jun 2021
16. Y. Cherdantseva, P. Burnap, A. Blyth, P. Eden, K. Jones, H. Soulsby, K. Stoddart, A review of cyber security risk assessment methods for SCADA systems. Comput. Secur. **56**, 1–27 (2016)
17. https://owasp.org/www-project-top-ten/. Accessed 15 Jun 2021

18. R. Leszczyna, Review of cybersecurity assessment methods: applicability perspective. Comput. Secur. 102376 (2021)
19. A. Kumari, R. Gupta, S. Tanwar, Amalgamation of blockchain and IoT for smart cities underlying 6G communication: a comprehensive review. Comput. Commun. (2021)
20. C. Schmitz, S. Pape, LiSRA: lightweight security risk assessment for decision support in information security. Comput. Secur. **90**, 101656 (2020)
21. H. Qiu, T. Dong, T. Zhang, J. Lu, G. Memmi, M. Qiu, Adversarial attacks against network intrusion detection in IoT systems. IEEE Internet Things J. (2020)
22. R. Knight, J.R. Nurse, A framework for effective corporate communication after cyber security incidents. Comput. Secur. **99**, 102036 (2020)
23. http://www.ignou.ac.in/. Accessed 25 Jun 2021
24. https://www.wireshark.org/download.html. Accessed 25 Jun 2021
25. A. Kumari, R. Gupta, S. Tanwar, N. Kumar, A taxonomy of blockchain-enabled softwarization for secure UAV network. Comput. Commun. **161**, 304–323 (2020)
26. K. Sanjeev, B. Janet, R. Eswari, Automated cyber threat intelligence generation from honeypot data, in *Inventive Communication and Computational Technologies* (Springer, Singapore, 2020), pp. 591–598
27. A. Kumari, S. Tanwar, Secure data analytics for smart grid systems in a sustainable smart city: challenges, solutions, and future directions. Sustain. Comput. Inf. Syst. **28**, 100427 (2020)
28. Y. Sun, Z. Tian, M. Li, S. Su, X. Du, M. Guizani, Honeypot identification in softwarized industrial cyber-physical systems. IEEE Trans. Industr. Inf. **17**(8), 5542–5551 (2020)
29. W. Tian, M. Du, X. Ji, G. Liu, Y. Dai, Z. Han, Honeypot detection strategy against advanced persistent threats in industrial internet of things: a prospect theoretic game. IEEE Internet Things J. (2021)
30. W. Tian, M. Du, X. Ji, G. Liu, Y. Dai, Z. Han, Contract-based incentive mechanisms for honeypot defense in advanced metering infrastructure. IEEE Trans. Smart Grid
31. A. Kumari, S. Tanwar, S. Tyagi, N. Kumar, Verification and validation techniques for streaming big data analytics in the internet of things environment. IET Netw. **8**(3), 155–163 (2018)
32. S.V.B. Rakas, M.D. Stojanović, J.D. Marković-Petrović, A review of research work on network-based SCADA intrusion detection systems. IEEE Access **8**, 93083–93108 (2020)

Chapter 9
Importance of Cyberlaw

1 Introduction

Gradually, the society has become Internet-dependent. Every individual and, therefore, the nation depend heavily on cyberspace for day-to-day activities. This is true for every nation in the world at the present moment. Keeping track of not only the information but also the activities of individuals over the Internet has become a challenge. There are various types of threats with respect to cyberspace. In fact, the threats are evolving, and new types of attacks are also surfacing now and then. Cybercrime has become a regular phenomenon and another dimension of challenges the human race needs to address. Cybercrimes happen due to massive and coordinated attacks against an information infrastructure. Such an information infrastructure may be confined to an organization, an institution, a company, or even to a nation. Cybercrimes may affect any individual who directly or indirectly interacts with the networked services. Such networked services may be available over the Internet or other networks owned by different business houses, enterprises, government departments, organizations, institutions, and so on so forth [1].

Security of information is a major concern in present society. Lack of security of information is the major reason behind cybercrimes. The concept of cybersecurity emphasizes protecting information, computers and various computer resources, communication infrastructure and communication devices, and protecting various digital devices and equipment. Information needs to be protected from unauthorized access, disclosure, modification, destruction, and unauthorized use.

Cybercrimes are evolving in their nature. Novel cyberattacks are arriving and also may be expected to arrive in the future. For example, credit card fraud, accessing bank accounts over the Internet, defaming people on online platforms such as social media, gaining unauthorized access to computers and data, identity theft (stealing the identity of another person) in order to do different criminal activities, violating copyright, trademarks, software license, software piracy, cyberstalking, child pornography in online platforms, etc., are some of the cybercrimes very much common in

N. Dutta et al., *Cyber Security: Issues and Current Trends*, Studies in Computational Intelligence 995, https://doi.org/10.1007/978-981-16-6597-4_9

present days. Persons involved in doing these activities are the cybercriminals. Cybercrime can happen against an individual; it can happen against property like credit card frauds, violating intellectual property rights in online platforms, etc. Similarly, cybercrime can happen against organizations, and it can also occur against society [2].

A few well-known cyberattacks and therefore cybercrimes are enlisted here.

E-mail spoofing: Here, an e-mail appears to have originated from a particular source, but in reality, the e-mail has been sent from another source.

Spamming: E-mail spam is a very common type of attack in the present scenario. Unsolicited e-mails arrive in the mail inbox.

Internet time theft: An authorized person may use the Internet time (i.e., hours) purchased by another person. It is a kind of hacking. Internet time theft kind of crime is conducted through identity theft.

Cyberdefamation: This happens when defamation takes place in the cyberworld. Cyberdefamation occurs in electronic form and with the help of computers, mobile phones, and the Internet. For example, if a person publishes some defamatory matter against another individual on a website, or social media, or through e-mails even, then it is a matter of cyberdefamation.

Salami attack: This is a crime related to the financial matter. Under this type of attack, the alterations made in the financial data are so insignificant that it generally goes unnoticed. For example, suppose a computer program is installed in the banking system that can deduct a very small amount of money from a large number of bank accounts (e.g., Rs.1 every month). In that case, the attacker may be successful in collecting a huge amount every month, and at the same time, the unauthorized transaction may go unnoticed by individual account holders.

Forgery: Postage and revenue stamps, currency notes, certificates, mark sheets, etc., are possible to forge by using a sophisticated computer, printer, scanners and software. Huge monetary scams are possible to happen through such forgery.

Web Jacking: This attack happens when an attacker forcefully takes control of a website. The original owner may not have any more control over the website once the attacker takes over it. This may happen if the attacker steals the administrative passwords.

This list is partial, and the complete list of cyberattacks may be very long. Thus, it is for sure that no one in the cyberspace is absolutely safe. The Internet revolution has also brought this dark side to the information society. The global information society which lives in the single Internet is very much vulnerable to the cyberattacks.

Creating solutions to the problems in the context of the cybercrime is a continuous ongoing process. Technology is an option to offer defense against cyberattacks. However, to maintain a peaceful ambiance in society and increase the trust level of the users of the Internet and computers, it is also necessary to have a legal system in place worldwide. It may be highly challenging to have one uniform legal system to address the cybercrimes across the world since every nation or geographical territory has its own legal system, but there has to be effort globally to address issues such as the users of the Internet, and the global information society living on the Internet does not obey geographical demarcations. In such a global society, crime may originate in one nation, and its effect may be visible in another nation.

In this chapter, various issues with respect to the legal frameworks to address the cybercrimes are highlighted. The necessity of cyberlaw is highlighted in Sect. 2. The global landscape of cyberlaw is presented in Sect. 3. Various types of cybercrimes and a discussion on cybercrimes that are happening across the world in the current time are presented in Sect. 4. The chapter is concluded in Sect. 5.

2 Why Cyberlaw is Necessary

Usage of computers and computer networks also invites risks. Several security-related issues arose due to the heavy use of computers and computer networks like the Internet. This security is linked with the privacy of the users, unauthorized access to information, eavesdropping to ongoing communications, data protection, financial and other transactions over the Internet, fraud, and many more. Moreover, issues in cyberspace like freedom of expression, securing intellectual properties, and other crimes are committed by using computers, and computer networks are major concerns in modern society. Cyberlaw is a framework that can be deployed in order to give legal recognition to all the risks and threats that can arise due to the usage of computers and computer networks.

Every nation has its own legal system. The legal framework of one nation cannot be applied in another nation, as it is a matter of jurisdiction. However, there has been effort worldwide to come up with international law, and it is a continuously evolving process. Along with the arrival of the Internet in society and across the globe, a new form of society has taken birth which we know as a cybersociety or information society or cyberworld (and so on so forth). Technology has always been like a double-edged sword. Inappropriate behavior of someone in cyberspace may invite trouble to others. Even there can be planned and organized activities by a group of people over the Internet, which can be a threat and undesired for some others. Thus, the Internet may be a place for criminal activities even. Interestingly, the Internet is one society that does not obey the law of geographic boundaries. Just by one mouse click, a citizen of one country can access a website that originated in another country. Therefore, in many situations, crime can happen in a different country, and the cause (the criminal) can be located in a different country. The issue of jurisdiction of law again surfaces in such situations. The types of crime or criminal activities over the Internet are novel and surely different from the practices that prevail in the non-Internet-based society. Moreover, a new type of crimes is also getting devised in recent times. In fact, it is highly important to envision the types of crimes that can happen in the society in the days to come. On the other hand, even if we avoid the discussion regarding the use of the Internet and other computer networks, we still see that the use of computers and similar digital devices in society is dominant. Such uses of computers and other similar devices, including mobile phones, have changed the overall behavioral pattern of society. Organizations, industries, institutions, departments, and households use computers heavily, and it seems that the use of computers and similar digital devices is indispensable and also critical to their survival. However, penetration of such devices in the society has also invited different criminal activities along with, although the crime patterns are novel.

The old-aged legal framework and practices prevailing in the nations definitely do not have provisions to address such new challenges concerning the crimes that can occur due to the use of computers and computer networks like the Internet. Those legal systems across different nations may be robust enough to handle other forms of crimes, but considering the recent arrival of cybercrimes in the society and their unique nature, the existing legal systems must provide mechanisms to address this new challenge.

Electronic commerce (e-commerce) is another advent that has penetrated into the society along with the growth of the Internet. The user base of e-commerce is expanding day by day. There can also be frauds in the e-commerce paradigm from different perspectives. Again e-commerce is a new form of business that takes place over the Internet. Thus, it needs an addendum to different existing business laws and tax laws across the nations, as commerce involving multiple nations is a very common practice in e-commerce.

As a summary, the following are the major reasons why cyberlaw is necessary.

- The nations may have established and well-defined respective legal systems that can address all possible situations and cases in the pre-Internet era. But due to the arrival of computers and the Internet, there is a need to have a legal framework that can address all the legal matters that can arrive due to the increasing use of computers, associate technologies, and the Internet. So, this gap needs to be bridged by suitable laws and an appropriate legal framework known as cyberlaw.
- Internet needs to be given legal recognition. Again due to the advent of social networks and e-commerce, the Internet has become one of the most dominating technologies across the world. Thus, Internet is a place where the entire world converges. And Internet perhaps is the most popular technology that was conceived in the twentieth century.
- Cloud platforms (which are high-end computing platforms generally owned by third parties) have become the major places for computing, data storage, and communications too. Considering the dependencies of users on cloud platforms, it is highly important to give legal recognition to the cloud technology as a whole.
- Cyberterrorism is now common. It includes disruptive activities centered on ideological, political, religious, or similar notions. This is an old offense as marked by the society, but due to the popular use of the Internet, it can be committed using innovative ways.
- At present, the society is driven by data. Internet is the major source of data, although not the only one. Thus, data has become a part of the society. Any disruption to such data will have an impact on the society. Therefore, data safety has become a significant concern.

3 Global Landscape of Cyberlaw

Crime or an offense may be stated as follows: "*it is a legal wrong that can be followed by criminal proceedings which may result into punishment.*" Criminality is nothing but is a breach of the law made for criminals and wrongdoings. In this section, an effort shall be made to draw the legal landscape across the globe with respect to cybercrime. *Cybercrime* and *privacy protection* are two essential dimensions of the cyberworld and the modern information society. In order to build an information society, the users need to trust electronic communication systems, computerized information processing systems, and the Internet as a whole. Lack of trust will discourage users from sharing information in the digital information society. Thus, *privacy protection* is an important dimension in the trust framework. Technology design, industrial practices, and legal framework may contribute significantly in assuring privacy to the users. However, considering the legal dimension, bringing cybercriminals under one legal framework is a challenge as there is no central authority. This subject is evolving rapidly as it is an urgent need in the digital global society. Enforcing uniform cyberlaw across the World Wide Web (WWW) irrespective of the citizenship of the cybercriminals may be a challenge as every land has its own legal systems.

Asia Pacific Region

There is a lack of awareness regarding various issues related to information security in the Asia Pacific region. Increasing complexity, capacity, and reach of ICT are some of the issues people need to be aware of. Presence of the communication networks in multiple nations is a fact people must be aware of before making use of the network. In fact, there is a possibility that multiple cybercrimes that are taking place every day may have gone unnoticed and unreported. Only a few countries in the Asia Pacific region have legal and regulatory frameworks to handle cybercrime-related issues.

Data protection is about creating such a trusted framework in which data collection, exchange of data, and personal data in commercial, governmental, and social contexts remain truly safe and secure.

Australian Cybercrime Act 2001 came into force in the year 2002. There are criticism about this act. It is said that the definition of cybercrime as per the act is too broad. Moreover, definitions of restricted data and authorization are two concerning aspects of the act. IT professionals must take serious care while performing their duties and be more aware to avoid the risk of prosecution for their otherwise well-intentioned actions.

Data privacy, spam, and online child safety are few major concerns concerning cybersecurity in broader sense. Identity theft may lead to data privacy, and similarly, unwanted mails are spam and are a form of cyberattack only.

There is no central regulation to handle such cybersecurity-related issues. However, in some particular domains such as computer security and online child safety, there are some international norms to decide on the best approach to regulation. For example, Children's Online Privacy Protection Act (COPA) may be referred to handle online child safety-related issues. In addition, the Council of Europe's

(CoE's) Convention on Cybercrime is a widely regarded international norm to handle computer crime. Similarly, the International Centre for Missing and Exploited Children (ICMEC) has developed authoritative model legislation to handle child pornography. This model is accepted as the benchmark legislation to handle computer security and child safety-related matters in the Internet. However, there is no international norms to handle areas like privacy and spam.

There are several regional norms to handle privacy issues, for example, the Asia–Pacific Economic Co-operation (APEC) Privacy Framework and the EU's Data Protection Directive. Unfortunately, there is a lack of international agreements on the topic, like how to address data protection issues. However, the CoE's Convention on Cybercrime provides the benchmark legislation.

APEC Privacy Framework provides nine principles: preventing harm, integrity of personal information, notice, security safeguards, collection limitations, access and correction, uses of personal information, accountability, and choice. Fort details on privacy, the book mentioned in (4) may be referred.

When we look at the necessities of law to deal with cyberattacks, we find the following four areas that draw maximum attention: Computer Security Law, Data Privacy and Data Protection Law, Spam Law, and Law of Online Protection for Children.

Unauthorized access is the major threat to computer security. Similarly, preparation, production, dissemination, and use of computer viruses and malware are also different forms of cybercrime. As mentioned earlier, CoE's Convention on Cybercrime provides the benchmark around which different countries in the Asia Pacific region have aligned their own laws and framed different respective acts.

Countries like Australia, New Zealand, Singapore, Taiwan, Thailand, China, Hong Kong, Japan, South Korea, Malaysia, the Philippines, and Vietnam have framed their computer security laws with different levels of alignments to CoE's Convention on Cybercrime. Similarly, India and Indonesia also have framed their own acts, although degree alignment with CoE's Convention on Cybercrime is week.

The Information Technology Act, 2000 (also known as ITA-2000, or the IT Act) is an Act proposed by the Government of India on October 17, 2000. It is the primary law in *India* that deals with *cybercrime* and issues related to *electronic commerce*. This law is based on the *UNCITRAL Model Law on International Commercial Arbitration* recommended by the United Nations General Assembly by a resolution dated January 30, 1997.

The laws apply to the whole of India. If a crime involves a computer or network located in India, persons of other nationalities can also be indicted under the law. The Act provides a legal framework even for electronic governance. It gives recognition to electronic records and digital signatures. It also defines different cybercrimes and also prescribes penalties for them. It also established a Cyber Appellate Tribunal to resolve any disputes arising from this new law.

Amendments to ITA-2000: A major amendment to this law was made in the year 2008. Under this amendment, provisions to address—pornography, child porn, cyberterrorism, and voyeurism, were introduced. The amendment was passed on December 22, 2008, without any debate in Lok Sabha. The next day it was passed by the Rajya Sabha. It was signed into law by President Pratibha Patil, on February 5, 2009.

Anti-Spam Law in Canada

Canadian Government proposed anti-spam legislation, Bill C-27, The Electronic Commerce Protection Act in the year 2009. This legislation addresses the issues that can arise due to spam, counterfeit websites, and Spyware. Amendments were brought to the Personal Information Protection and Electronic Documents Act to cover online privacy in detail and guidelines for e-mail marketing.

Canada has two different federal privacy laws, namely the Privacy Act and Personal Information Protection and Electronic Documents Act (PIPEDA).

There are two laws proposed by Canadian government, namely Senate Bill S-220 and Parliamentary Bill C-27 in the year 2009. The Senate Bill S-220 is said to be the Anti-Spam Act. This bill allows Internet Service Providers to refuse, filter, and even block e-mails which are actually spams. This bill also considers phishing attacks. On the other hand, Parliamentary Bill C-27 is an Anti-Spam Act. It covers issues like e-mail communications with business intention without permission of the receiver, unauthorized installed application programs, and alteration of data while under transmission between the sender and the receiver.

Federal Laws in USA

The United States House of Representative approved H.R. 5938 in the year 2008. This bill enables Federal Government to prosecute cybercriminals for identity theft. The bill allows victims to have claims for compensation. This bill has provisions for monetary fine and imprisonment up to a duration of 5 years for Spyware. Criminal penalty may be imposed on use of malicious Spyware. If proved guilty, accused criminals will have to pay a monetary fine and undergo imprisonment up to 1 year. According to this bill, obtaining, deleting, or releasing data from a computer is considered to be a crime. To threaten to crash a computer is also considered to be a criminal activity. Cyberextortion (demanding money against a protected computer) is considered to be crime as per this bill. Cyberfraud and making intentional false representation online are offences in the USA. Again, unauthorized use of the social security number of another person, credit card information, driving license information, and work identity number (ID) are cybercrimes as per US federal laws.

EU Legal Framework

The European Union (EU) legal framework addresses information management principles in fairness, transparency, consent, purpose specification, data retention, security, and access. A highly developed area of law in Europe is the right to privacy. It is believed in the EU that law enables trust and confidence in the information society. Data Protection Directive, also known as the EU directive, is the most important part of EU privacy and human rights law. This directive regulates the processing of personal data within the EU. European Commission implemented EU directive in the year 1995.

CoE's Convention on Cybercrime (2001) provides the basis of cyberlaw EU. Following activities are criminalized as per CoE's Convention on Cybercrime:

Illegal access to computer systems,
Interfering with the computer system without right,
Intentional interference with computer data without rights,
Illegal interception of data to a computer system,
Forgery of data,
Infringement of copyright-related rights online,
Child pornography related offences.

There are similarities between regulations of USA and EU concerning law enforcement against cybercrime.

Cyberlaw in African Region

Different members of African Union have adopted legal framework for cybercrimes. For example, Mauritius, South Africa and Zambia have adopted legislations for handling cybercrime. Similarly, Botswana and Gambia have also adopted legislations. On the other hand, in East Africa region, Tanzania, Kenya, and Uganda have adopted such legal framework to deal with cybercrimes. The countries like Algeria, Nigeria, Ghana, and Niger have adopted such legal framework. Although little slow, gradually, the entire African region has started the process of strengthening of legislations to handle cybercrimes.

4   Cybercrimes

In the present moment, the world is highly connected. But, importantly, it is seen that the tendency to remain connected has increased significantly in the society. This is because we are living in a technologically advanced society in which sophisticated technologies are emerging continuously.

Seamless connectivity is a desire everyone has in the society. This digital transformation has shown path toward a more cohesive and connected society. Data is being generated enormously over the Internet, and our data is now shared in many platforms. The emergence of data centers, cloud platforms, and also Internet of Things (IoT) have changed our life styles. Such a trend shall be growing only now onward.

However, such developments have come at a price. The society has become more connected than ever, and at the same time, our data has become more vulnerable.

The emergence of cyberworld has also given birth to cybercrimes. Under the category of cybercrime, the computer plays the role of the object of crime. Rather it is sued as a tool to commit an offense. Cybercriminals use digital devices to gain access to other's personal information. They try to gain access to confidential government data, business data or institutional data. Cybercriminals even try to disable other's digital devices connected to the Internet. They try to earn money through different fraudulent activities including selling of other's sensitive data.

Cybercrimes are increasing at a high pace across the world. Due to which companies, organization, and even individuals are losing huge amount of money. In fact, it seems, there is no end to it. Law enforcement is highly important and needs of the hour. Law enforcement may help in the attempt to tackle the issues related to cybercrime. It is essential to have laws in place in order to control the cybercriminals which is growing in numbers every year.

Experts Views on the Trends of Cybercrimes

Experts put forward their views regarding how the changing landscape of cybercrime is in the present scenario. According to the experts, the following are the major dimensions of cybercrimes, and the society shall have to face the year 2020 onward.

Uncontrolled Access to Personal Data: Due to the heavy use of online platforms, personal data shall be available in multiple platforms. There is high chance of data leakage and stealing of such data, and therefore, the digital society may have to face the risk of getting destabilized.

Smart Consumer Devices: Different smart devices including smartphones are getting penetrated into the society in a big way. However, the security risks are not analyzed in the way it should have been. This is because devices are getting launched and also adopted by the society before analyzing security risks and designing appropriate security solutions.

Risks of Using Medical Devices: Different medical rather bio-medical devices are being made available for use. Such devices usually are connected to the Internet. Therefore, there shall be threats to the use of these devices, and health crisis may have to be faced by the society due to such connectivity.

Challenges to the Vehicles and Transport Infrastructure: Attackers are going to target vehicles and transport infrastructures. Cyberattack is going to be very common in these infrastructures. The more connectivity we look for more challenges we shall have to face.

Smart Supply Chain are New Targets: Attackers shall target supply chains. Even software supply chain is going to face tremendous challenges from security perspectives.

Such cyberattacks shall have a huge impact on the economy and may have a very large range of affected victims.

Threats to Shipping: The ports and ship network are going to face substantial cybersecurity threats. Ship communication networks are already under attack. This shall increase in the days to come.

Vulnerabilities in Real-Time Operating Systems (RTOS): The real-time operating systems installed in different smart devices and embedded systems will be attacked severely. Cyberattack is going to be very common on RTOS.

4.1 Categories of Cybercrime

In general there are three categories of cybercrime from the impact area perspective. The categories are "individual", in which an individual is affected; "property", in which the property like a bank account of a person goes to the control of cyberattacker; and "government", where attackers gain access to the website or digital infrastructure of the government.

4.2 Types of Cybercrime

From a technical perspective, the following are different types of cybercrimes that are taking place in around and seen very often: distributed denial of service (DDoS) attacks, botnets, identity theft, cyberstalking, social engineering, potentially unwanted programs, phishing, prohibited/illegal content sharing, exploit kits, and online scams.

Cyberoffences Enlisted

In this section, few cyberoffences are enlisted. However, such offences may occur in any nation, and therefore, the nation needs to frame a legal framework to address these issues. The following table containing list of different offences is prepared as per the ITA 2000 (Government of India) [3]. Therefore, it is worth mentioning that the sentences are also borrowed from the act.

Offence	Description
Tampering with documents stored in computer	This indicates knowingly or intentionally concealing, destroying, or altering or intentionally or knowingly causing another to conceal, destroy, or alter any computer source code used for a computer, computer program, computer system, or computer network when the computer source code is required to be kept or maintained by law for the time being in force
Hacking the computer system	This indicates if a person with the intent to cause or knowing that he is likely to cause wrongful loss or damage to the public or to any person, destroys or deletes or alters any information residing in a computer resource or diminishes its value or utility or affects it injuriously by any means by, commiting hack

(continued)

(continued)

Offence	Description
Receiving stolen computer or device used for communication	This indicates if a person receives or retains a computer resource or communication device which is known to be stolen or the person has reason to believe is stolen (One should never receive any stolen item that may be offered at lower prices, it does not matter; it is a crime.)
Using the password of another person	This indicates if a person fraudulently uses another person's password, digital signature, or other unique identification
Cheating through computer resource	This indicates if a person cheats someone using a computer resource or communication
Publishing private images of others	This indicates if a person captures, transmits, or publishes images of a person's private parts without their consent or knowledge
Acts of cyberterrorism	This indicates if a person denies access of authorized personnel to a computer resource, accesses a protected system, or introduces contaminant into a system, intending to threaten India's unity, integrity, sovereignty, or security, then he commits cyberterrorism
Publishing obscene information (in electronic form)	This indicates if a person publishes or transmits or causes to be published in the electronic form, any material which is lascivious or appeals to the prurient interest or if its effect is such as to tend to deprave and corrupt persons who are likely, having regard to all relevant circumstances, to read, see, or hear the matter contained or embodied in it
Publishing images containing sexual acts	This indicates if a person publishes or transmits images containing a sexual explicit act or conduct
Publishing porn involving a child or predating children online	This indicates if a person captures, publishes or transmits images of a child in a sexually explicit act or conduct. If a person induces a child into a sexual act. A child is defined as anyone under 18
Failure to maintain records	This enforces that persons deemed as intermediaries (such as an ISP) must maintain required records for a stipulated time. Failure is an offense

(continued)

(continued)

Offence	Description
Failure/refusal to comply with orders	This indicates the Controller may, by order, direct a Certifying Authority or any employee of such Authority to take such measures or cease carrying on such activities as specified in the order if those are necessary to ensure compliance with the provisions of the act, rules or any regulations made thereunder. Any person who fails to comply with any such order shall be guilty of an offense
Failure/refusal to decrypt data	This indicates if the Controller is satisfied that it is necessary or expedient so to do in the interest of the sovereignty or integrity of a State, the security of the State, friendly relations with foreign states or public order or for preventing incitement to the commission of any cognizable offence, for reasons to be recorded in writing, by order, direct any agency of the Government to intercept any information transmitted through any computer resource. The subscriber or any person in charge of the computer resource shall when called upon by any agency which has been directed extend all facilities and technical assistance to decrypt the information. The subscriber or any person who fails to assist the agency referred is deemed to have committed a crime
Securing access or attempting to secure access to a protected system	This indicates that the appropriate Government may, by order in writing, authorize the persons who are authorized to access protected systems. If a person who secures access or attempts to secure access to a protected system, then he is committing an offence
Misrepresentation	This indicates if anyone makes any misrepresentation to, or suppresses any material fact from, the Controller or the Certifying Authority for obtaining any license or Digital Signature Certificate

Trends in Cybercrimes

One very important aspect of framing cyberlaw is that one has to understand the inner details of a cyberthreat or attack or crime before framing the law to convict the criminals. Thus, lawmakers are expected to be technocrats as well to handle cybercrimes. Therefore, in this section, a few cyberattacks that took place in recent years are highlighted. This will make the cybercrime landscape clear and also help in understanding the trends.

Due to increasing awareness in the area of cybersecurity and deployment of security solutions by the users, it is also challenging for the attackers to continue with same attack again and again. Therefore, the attackers continuously search for vulnerable areas. In this section, some of the common attacks that took place in the year 2019 worldwide are enlisted.

Software Supply Chain Attack

Under this attack, an attacker installs malicious code in legitimate software. The attacker does this in one of the building blocks of a particular software relies upon. There are two main categories of software supply chain attack. These are targeted attack and locating weak link. Under targeted attack, attacker aims to compromise well-defined targets and scans their suppliers list in search of the weakest link through which they can enter. For example, the **ShadowHammer** attack on ASUS is an example of a targeted attack. Attackers considered ASUS live update utility and implanted malicious code there. Thus, attackers could establish a backdoor to millions of computers remotely located. Under the second category of this attack (i.e., locating a weak link), the attacker tries to locate a weak link with a large distribution radius. With this weak link, the attacker tries to compromise as many victims as they can. An example of this category of attack is the attack on **PrismWeb** (an e-commerce platform) in which attackers considered shared JavaScript libraries. A skimming script was injected in these libraries through which more than two hundred online stores in North America were affected.

There has been a sharp increase in the number of software supply chain attack in the USA. It is a growing trend. The US government's Home Land Security department has established Information and Communications Technology Supply Chain Risk Management Task Force (2019). The government of the United States has declared foreign supply chain threats as a national emergency. The government has also empowered the Secretary of Commerce to prohibit transactions having doubts. Such an order led to the decision to ban the technology giant Huawei. The magnitude and severity of such an attack are a concern. This type of attack is very dangerous, and they strike at the supplier–customer relation's trust.

E-mail Scams

E-mail scams are very common. In fact, every individual and organization are exposed to this attack multiple times. But with the growing public awareness about e-mail attacks, the attackers have improved phishing tactics. Attackers try to establish credibility among the victims by following smart approaches. They also use advanced evasion techniques to break the e-mail security solutions. Sextortion scams and Business E-mail Compromise (BEC) are two major forms of attacks. Through sextortion, attackers trick victims and blackmail seeking monetary payment. In BEC, attacker convincingly impersonates others. In modern form of the e-mail scams, it is not necessary to have any malicious attachments or links. This makes such attacks even harder to detect. The attackers apply social engineering techniques. The attackers are successful in personalizing the e-mails' content and fly safely under the radar of anti-spam filters. Finally, they reach the inbox of their target.

Attacks Against Cloud Environment

The public cloud environments are getting popularity. Considering the available cloud service provider in the market, enterprises are migrating their storage and computing infrastructure to them increasingly. But the cloud environments need to be more safer. Cloud environment providers must deploy best practices known so far to protect their environment. As a whole, misconfiguration and not so strong management of cloud resources are the major concerns considering the security breaches that the cloud environment may suffer from. Cloud assets are subjected to a wider array of attacks. Amazon cloud servers and Microsoft cloud servers are also in the list of attacked cloud platforms in the year 2019. Public cloud infrastructures are being targeted in an increasing manner. Apart from information theft, attackers even intentionally abuse different cloud technologies for their computing power.

Mobile Landscape

These days, personal and business or work lives are heavily dependent on mobile phones and other such devices connected to the Internet. Large number of sensitive data are generally stored in these mobile devices. This fact encourages attackers to devise and launch different attacks targeting these devices spread across. For example, profitable advertising campaigns, sensitive credential theft using fake apps, surveillance operations, etc., are already launched attacks. More and more such types of general attacks are nowadays translated to the world of mobile.

Banking malware is widely deployed in the mobile arena also. Malware that can steal payment data, user credentials, and even funds from the bank accounts of victims has become a great mobile threat.

Targeted Ransomware

Targeted ransomware is still an active approach used by attackers. Government entities, corporate networks, airport networks, cloud hosting providers, etc., are targets for the attackers.

Cryptominers

Cryptominers are a malware type that is still prevalent. They target corporations, factories, server platforms, and even cloud platforms. Moreover, attackers are integrating cryptominers as a part of botnet to launch distributed denial-of-service attacks.

DNS Attack

Domain Name System (DNS) is a highly important and successful mechanism used in the Internet. The DNS is used to resolve or translate domain names into corresponding Internet Protocol (IP) addresses. DNS is an important ingredient in the Internet trust framework. Attackers try to compromise DNS providers, name registrars, and local DNS servers belonging to the targeted organization. Attackers try to manipulate the DNS records. If attackers are successful in taking over the DNS, the whole network can be compromised. Such a takeover will enable attackers to launch multiple attacks

Table 1 Categories of cyberattack in different regions

Category	Global (%)	Americas (%)	EMEA (%)	APAC (%)
Mobile	30	39	25	35
Cryptominers	21	26	19	23
Botnet	13	16	7	16
Banking	6	8	4	10
Ransomware	3	3	2	3

EMEA Europe, the Middle East & Africa
APAC Asia Pacific

Table 2 Malicious file types under HTTP

Under the protocol	xls	js	msi	doc	pdf	exe	Others
HTTP (%)	2	2	3	5	6	53	29

Table 3 Malicious file types under SMTP

docx	xlsx	pdf	rtf	js	doc	exe	Others
3%	8%	11%	12%	16%	19%	21%	10%

in the network, e.g., controlling e-mail communications, redirecting the victim users to different phishing sites, etc.

DNS attacks are growing trends. It is a significant risk in the entire Internet infrastructure. There are a large number of incidents that involve DNS attacks.

Statistics of Cyberattacks During the First Half (H1) of 2019

The following table demonstrates different categories of cyberattacks that took place in different regions of the Globe during 2019 (first half, JAN-JUN) [4] (Tables 1, 2, and 3).

5 Conclusion

The growing trend of cybercrimes is a major concern in present-day society. There have been novel cybercrimes emerging every year. However, the legal frameworks of the nations are yet to be ready to handle all sorts of issues related to cybercrime. Some nations have systems in place to handle such issues, although evolving. Ultimately, to have a stable digital society, there is an urgent need for a robust legal system worldwide. International laws need to be framed, and there is a need for consensus among the nations. In this chapter, the legal landscape across the world with respect to cybercrimes has been depicted concisely. Various types of cybercrimes that are taking place are enlisted. Current trends in the cybercrimes area are elaborated. It is emphasized that legal systems are required to be in place with international consensus to handle ever-growing cybercrimes across the world.

References

1. V. Sharma, *Information Technology Law and Practice*, 4th edn. (Universal Law Publishing, 2015). ISBN 978-93-5035-527-5
2. N. Godbole, S. Belapure, *Cyber Security* (Wiley, 2016). ISBN 978-81-265-2179-1
3. The Information Technology Act 2000, accessed in https://www.indiacode.nic.in/bitstream/123456789/13116/1/it_act_2000_updated.pdf
4. Cyber Attack Trends: 2019 Mid-Year Report, Checkpoint Software Technologies Limited, accessed in https://research.checkpoint.com/2019/cyber-attack-trends-2019-mid-year-report/